Springer Theses

Recognizing Outstanding Ph.D. Research

Aims and Scope

The series "Springer Theses" brings together a selection of the very best Ph.D. theses from around the world and across the physical sciences. Nominated and endorsed by two recognized specialists, each published volume has been selected for its scientific excellence and the high impact of its contents for the pertinent field of research. For greater accessibility to non-specialists, the published versions include an extended introduction, as well as a foreword by the student's supervisor explaining the special relevance of the work for the field. As a whole, the series will provide a valuable resource both for newcomers to the research fields described, and for other scientists seeking detailed background information on special questions. Finally, it provides an accredited documentation of the valuable contributions made by today's younger generation of scientists.

Theses are accepted into the series by invited nomination only and must fulfill all of the following criteria

- They must be written in good English.
- The topic should fall within the confines of Chemistry, Physics, Earth Sciences and related interdisciplinary fields such as Materials, Nanoscience, Chemical Engineering, Complex Systems and Biophysics.
- The work reported in the thesis must represent a significant scientific advance.
- If the thesis includes previously published material, permission to reproduce this must be gained from the respective copyright holder.
- They must have been examined and passed during the 12 months prior to nomination.
- Each thesis should include a foreword by the supervisor outlining the significance of its content.
- The theses should have a clearly defined structure including an introduction accessible to scientists not expert in that particular field.

Rui Kamada

Tetramer Stability and Functional Regulation of Tumor Suppressor Protein p53

Doctoral Thesis accepted by
Hokkaido University, Japan

 Springer

Author
Dr. Rui Kamada
National Institute of Child Health
& Human Development
National Institutes of Health
Building 6, Room 2A11
6 Center Drive
Bethesda, MD 20892
USA

Supervisor
Prof. Kazuyasu Sakaguchi
Department of Chemistry
Faculty of Science
Hokkaido University
Hokkaido
Japan

ISSN 2190-5053   ISSN 2190-5061   (electronic)
ISBN 978-4-431-54134-9   ISBN 978-4-431-54135-6   (eBook)
DOI 10.1007/978-4-431-54135-6
Springer Tokyo Heidelberg New York Dordrecht London

Library of Congress Control Number: 2012939122

Printed on acid-free paper

Springer is part of Springer Science+Business Media (www.springer.com)

Parts of this thesis have been published in the following journal articles:

1. Kamada, R., Nomura, T., Anderson, C. W., Sakaguchi, K. (2011) Cancer-associated p53 tetramerization domain mutants: quantitative analysis reveals a low threshold for tumor suppressor inactivation, *J. Biol. Chem.* **286**, 252–258.

2. Kamada, R., Yoshino, W., Nomura, T., Chuman, Y., Imagawa, T., Suzuki, T., Sakaguchi, K. (2010) Enhancement of transcriptional activity of mutant p53 tumor suppressor protein through stabilization of tetramer formation by calix[6]arene derivatives. *Bioorg. Med. Chem. Lett.* **20**, 4412–4415.

Parts of this thesis have been published in the following journal articles:

Supervisor's Foreword

To understand the regulatory mechanism of a biological system, it is important to clarify the equilibrium of protein–protein interactions and the threshold for the signal output of cells in response to input cellular stimuli and stresses. Quantitative analysis is essential for truly understanding the equilibrium and threshold. In this thesis, we propose a novel mechanism of regulation of protein function, from the viewpoint of quantitative analysis of structural stability and protein function.

Using tumor suppressor protein p53 as a model protein, we demonstrated that the threshold for loss of tumor suppressor activity, in terms of the disruption of p53's tetrameric structure, could be extremely low. Our model revealed that p53 could respond quickly to stimuli by activating and regulating its function via small alterations in its tetrameric status. This is a novel and important discovery in the understanding of functional regulation mechanisms of proteins that play essential roles in biological systems.

Furthermore, we developed two projects concerning the functional control of p53 and reported a mutant p53 stabilizer and p53 inhibitor. We reported, for the first time, enhancement of the in vivo transcriptional activity of the most common Li-Fraumeni p53 mutant, R337H, by a calixarene derivative that stabilized the tetramer formation. To develop an efficient method of generating induced pluripotent stem cells, we demonstrated p53 inhibition via hetero-oligomerization with a tetramerization domain-derived peptide.

The results of the quantitative analysis presented in this thesis suggest that p53 is inactivated by amplification of destabilization in cells. Moreover, further studies should be directed toward the development of p53 stabilizers and inhibitors as new types of drugs.

Hokkaido, Japan, February 2012

Prof. Kazuyasu Sakaguchi

Acknowledgments

I would like to express my sincere gratitude to my supervisor, Professor Kazuyasu Sakaguchi, for his superior guidance, considerable encouragement, and invaluable discussion.

My heartfelt thanks go to Professor Takanori Suzuki (Laboratory of Organic Chemistry I, Hokkaido University) for his valuable suggestions and guidance on my papers, especially in organic synthesis. I would also like to thank Professor Yota Murakami (Laboratory of Bioorganic Chemistry) and Dr. Toshiaki Imagawa, Associate Professor, who gave me helpful and constructive suggestions and comments throughout this study.

I am very grateful to those who collaborated with me in this work: Dr. Carl W. Anderson (Brookhaven National Laboratory, USA), who provided me with invaluable comments and discussion; Professor Koichiro Ishimori (Laboratory of Structural Chemistry, Hokkaido); Professor Keiji Tanino (Laboratory of Organic Chemistry II, Hokkaido); Professor James G. Omichinski (Université de Montréal, Canada); Professor Yuji Kobayashi (Osaka Pharmaceutical University, Osaka); and Dr. Koichiro Fukuda (Laboratory of Organic Chemistry II, Hokkaido).

I wish to express my warm and sincere thanks to Dr. Yoshiro Chuman, Assistant Professor, who gave me helpful comments and encouragement. I also wish to thank the following people: Mr. Wataru Yoshino, for his essential assistance in organic synthesis; Dr. Takao Nomura, for his excellent advice and kind support throughout this study; Dr. Shunji Kaya, for his helpful advice and suggestions; and all members of the Laboratory of Biological Chemistry. In addition, I would like to thank the JSPS, Scholarship and Grant for Inter-laboratory Research Program, "Initiatives for Attractive Education in Graduate Schools", and the Hokkaido University Global COE Program for making my Ph.D. study possible with their financial support.

Finally, I extend my indebtedness to my parents for their support, understanding, and encouragement throughout my study.

17 February 2012

Rui Kamada

Contents

Abbreviations

ADC	Adrenocortical carcinoma
BD	C-terminal basic domain
CD	Circular dichroism
C_p	Het capacity
DAPI	4,6-diamidino-2-phenylindole
DBD	DNA binding domain
DCM	Dichloromethane
DIEA	N,N-diisopropylethylamine
DMF	Dimethylformamide
EGFP	Enhanced green fluorescence protein
FBS	Fetal bovine serum
F_{moc}	9-Fluorenylmethoxycarbonyl
f_u	Fraction of unfolded
HOBt	N-Hydroxybenzotriazole
HPLC	High performance liquid chromatography
HSQC	Heteronuclear single quantum coherence
iPS	Induced pluripotent stem
IPTG	Isopropyl β-D-1-thiogalactopyranoside
K_d	Dissociation thrmodynamic constant
MALDI	Matrix-assisted laser desorption/ionization
TOF-MS	Time of flying mass spectrometry
MDM2	Murine double minute protein
NHS	N-hydroxysulfosuccinimide
NLS	Nuclear localization signal
NMR	Nuclear magnetic resonance
OPTI-MEM	Reduced serum modified Eagle's medium
p53tet	p53(324–358)
p53TD	p53(319–358)
PBS	Phosphate buffered saline
PDB	Protein databank
PRD	Proline rich domain

PTD	Protein transduction domain
RPMI	Roswell Park Memorial Institute
SNP	Single nucleotide polymorphism
TAD	Transactivation domain
TD	Tetramerization domain
TFA	Trifluoroacetic acid
T_m	Melting temperature
WT	Wild-type

Chapter 1
General Introduction

1.1 Tumor Suppressor Protein p53

Tumor suppressor protein p53 plays a central role in maintaining genomic integrity [1–3]. In unstimulated cells p53 is maintained at very low levels but in response to DNA damage or other stress stimuli, such as hypoxia or activation of oncogenes, p53 becomes stabilized and accumulates in the cell [4, 5]. p53 can exert its tumor suppressor function in different ways (Fig. 1.1). It can function as a transcription factor that binds to the promoters of many target genes, such as *p21*, *mdm2*, *puma* and *bax*, thereby elevating or repressing their expression levels to induce cell cycle arrest and apoptosis. Nuclear and cytoplasmic p53 also physically interact with many other proteins to promote apoptosis and other processes, such as homologous recombination. In mitochondria, p53 contributes to transcription-independent apoptosis [6]. Moreover, in response to DNA damage, p53 enhances the post-transcriptional maturation of several microRNAs that have growth-suppressive functions [7]. Because p53 plays a prominent role in tumor suppression, it is a key target protein in cancer therapy. Recently, however, inactivation of p53 has also received attention because heart failure is associated with up-regulation of p53 function [8]. In addition, temporary suppression of p53 has been suggested as an approach to reduce side effects of cancer treatment and to increase the efficiency of iPS (induced pluripotent stem) cell generation, which is inhibited by p53–p21 pathways [9, 10].

1.2 Primary Structure of p53 Protein

p53 consists of five main domains: the N-terminal transactivation domain, the Pro-rich domain, the central DNA binding domain, the tetramerization domain and the C-terminal basic domain (Fig. 1.2). The N-terminal transactivation domain,

R. Kamada, *Tetramer Stability and Functional Regulation of Tumor Suppressor Protein p53*, Springer Theses, DOI: 10.1007/978-4-431-54135-6_1, © Springer Japan 2012

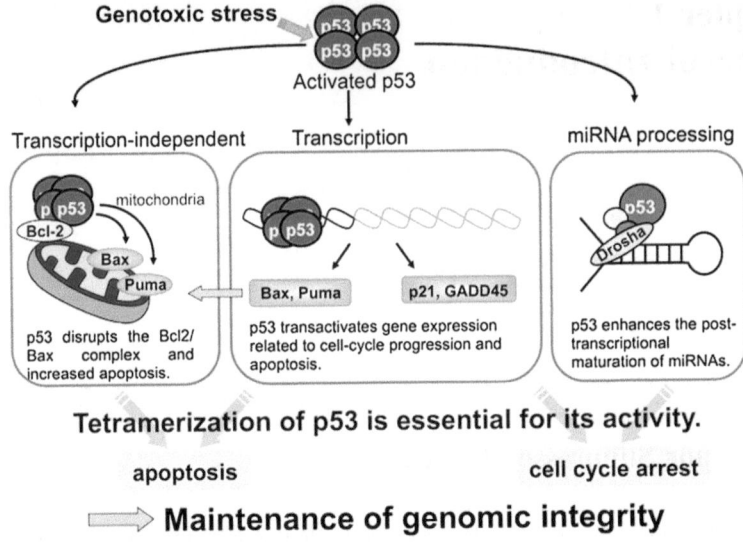

Fig. 1.1 Tumor suppressor protein p53. p53 becomes activated, stabilized and forms tetramers in response to genotoxic stress. Activated p53 suppresses tumorigenesis by inducing cell cycle arrest and apoptosis in different ways. 1. p53 induces apoptosis in a transcription-independent manner in mitochondria. 2. p53 transactivates or represses gene expression, leading to cell cycle arrest and apoptosis. 3. p53 enhances the maturation of microRNAs in response to DNA damage

residues 1–42, is required for transactivation activity and interacts with many proteins, including the transcription factors TFIID, TFIIH and several TAFs. The transactivation domain also mediates interaction with the histone acetyl-transferases CBP/p300 and PCAF and with the MDM2 E3 ubiquitin ligase. The Pro-rich domain, residues 61–92, is required for interaction with the Sin3 cor-epressor and is important for p53 stability, transactivation and for the induction of transcription independent apoptosis [11]. The central DNA binding domain can directly bind to the p53 consensus DNA binding site, which is composed of four pentanucleotide repeats. Full-length p53 can reversibly form tetramers via the tetramerization domain, residues 326–356. The tetramerization domain regulates the oligomeric state of p53. The C-terminal basic domain interacts with a non-specific DNA sequence and negatively regulates the sequence-specific binding of the central DNA binding domain [12, 13]. Deletion of the C-terminal basic domain increases sequence-specific DNA binding [14].

1.3 Posttranslational Modifications of p53

p53 can be modified by as many as 50 individual post-translational modifications, including phosphorylation, acetylation, mono- and di-methylation, glycosylation, ubiquitylation, neddylation, sumoylation and poly-ribosylation (Fig. 1.3).

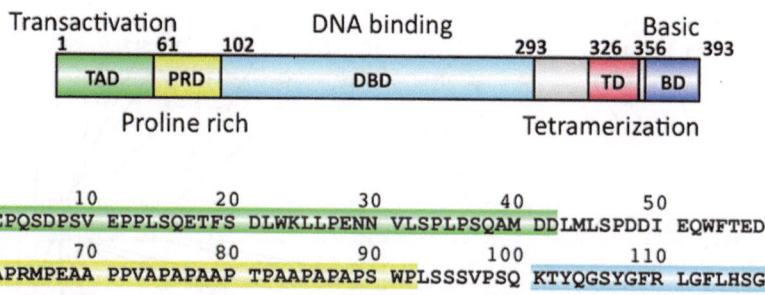

Fig. 1.2 Schematic structure and amino acid sequence of p53. p53 consists of five main domains: the N-terminal transactivation domain (TAD; 1–42, *green*), the Pro-rich domain (PRD; 61–92, *yellow*), the central DNA binding domain (DBD; 101–300, *light blue*), the tetramerization domain (TD; 326–356, *magenta*), and the C-terminal basic domain (BD; 364–393, *dark blue*)

These modifications, which occur mainly in the N- and C-terminal regions, regulate p53 activity and also tetramerization [15–17]. The N-terminal transactivation domain contains seven serines (Ser6, 9, 15, 20, 33, 37 and 46) and two threonines (Thr18 and 81) that are phosphorylated or dephosphorylated in response to ionizing radiation or UV light [2, 18–23]. p53 is maintained at low levels in normal, unstressed cells mainly through the action of MDM2, a RING finger type E3 ligase that promotes the poly-ubiquitylation and proteasomal degradation of p53. The main p53 targets of MDM2-mediated ubiquitylation are the six carboxy-terminal lysines (Lys370, 372, 373, 382, and 386). Phosphorylation of p53 Thr18, which requires prior phosphorylation of Ser15, induces dissociation of MDM2 from p53, resulting in p53 stabilization [24]. Phosphorylation of Ser46 in response to severe DNA damage enhances p53 mediated apoptosis by inducing the expression of p53AIP1, a protein associated with mitochondrial membranes that induces the release of cytochrome c to initiate apoptosis. Phosphorylation at Ser15, Thr18, and Ser20, in the N-terminal transactivation domain, induces conformational changes and switches the transactivation domain to more open conformations, which can interact with transcription factors, leading to enhancement of p53 transcriptional activity [25, 26]. Phosphorylation at Ser15 and Ser37 enhances p53-TFIID interaction and blocks MDM2 binding [27]. In contrast, phosphorylation of Ser15 alone inhibits p53-TFIID interactions.

In the C-terminal basic domain, Ser313, 314, 315, 366, 376, 378 and 392 and Thr377 and 387 can be phosphorylated, while Lys320, 372, 373, 371 and 382 can

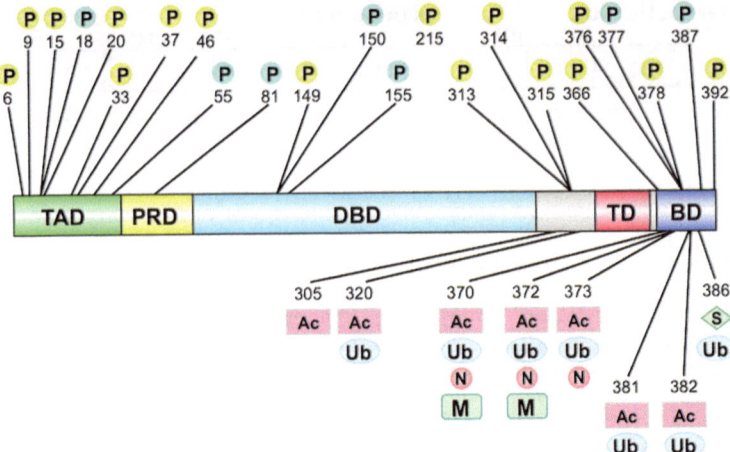

Fig. 1.3 Post-translational modification sites of human p53. Post-translational modifications occur mainly in the N- and C-terminal regions to regulate p53 activity. P (*yellow*), phosphorylation of Ser residue; P (*cyan*), phosphorylation of Thr residue; Ac, acetylation; Ub, ubiquitination; N, neddylation; M, methylation; S, sumoylation

be acetylated and Lys370, 372 and 382 can be methylated [28]. C-terminal phosphorylation and acetylation also promote p53 tetramerization and binding to PTEN, a dual specificity phosphatase [29–31]. Over-expression of the histone deacetylases HDAC-1, -2, or -3, or hSir2, which deacetylate p53, result in the transcriptional inhibition of p53 target genes; therefore, acetylation of p53 is important for its transcriptional activation. SMAR1 interacts with Ser15 phosphorylated p53 and MDM2 and recruits HDAC1 to p53, leading to the deacetylation of p53 and to decreased DNA binding [32]. Three C-terminal p53 residues, Lys370, Lys372, and Lys382 can be monomethylated and these modifications also modulate p53 activity in response to DNA damage. The methylation at Lys372 rapidly increases after DNA damage and promotes nuclear localization and increased stability of the p53 protein. On the other hand, methylation of K370 blocks p53 binding to DNA and represses transcriptional activation.

1.4 3D Structure of p53

The N-terminal region (TAD and PRD) and the C-terminal region (BD) are unfolded in their native states. However, when TAD binds to partner proteins, such as MDM2 and TFIIH, it folds into an amphipathic α-helical structure [33, 34]. The BD also changes its disordered structure to a folded structure upon binding to other proteins or nonspecific DNA [35]. In contrast to the unstructured N- and C-terminal regions, the central DNA binding domain and the tetramerization domain exist as folded structures. The DNA binding core domain, residues 101–300,

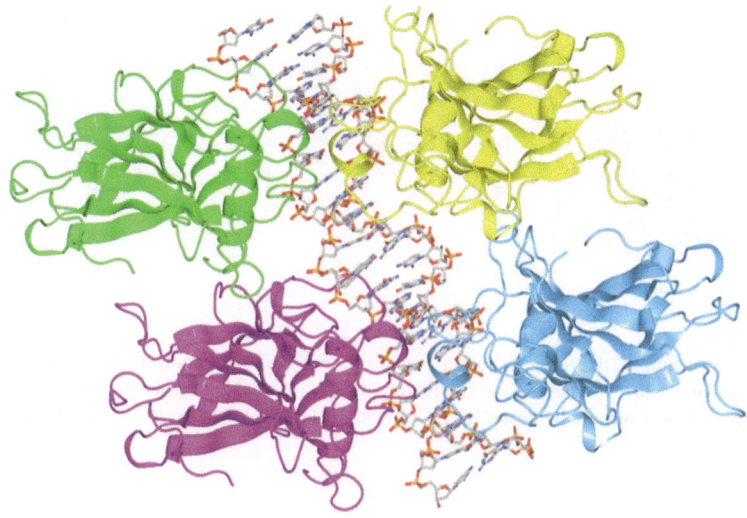

Fig. 1.4 Structure of the tetrameric p53DBD/DNA complex (pdb; 3EXJ). The four subunits of p53DBD are represented by *green*, *yellow*, *magenta*, and *cyan*. The p53DBD consists of a central immunoglobulin-like β-sandwich

consists of a central immunoglobulin-like β-sandwich that provides a scaffold for the DNA binding surface (Fig. 1.4) [36]. This surface is formed by two large loops, L2 an L3, that are stabilized by a zinc ion and a loop-sheet helix motif. The tetramerization domain, residues 326-356, exhibits dihedral symmetry and is described as a dimer of dimers (Fig. 1.5) [37–43]. Each monomer is comprised of a β-strand, a tight turn and an α-helix. Two monomers form a dimer via anti-parallel interaction of the β-strands and two dimers interact via a four-helix bundle to form a tetramer. The folding-unfolding process of the p53 tetramerization domain is in equilibrium between unfolded monomers and the folded tetramer [44]. Recently, the quaternary structure of full-length p53 has been revealed through electron microscopy, biophysical methods and computational modeling [45]. The model for p53 in solution, in the absence of DNA, is an elongated cross-shaped tetramer with extended N- and C-termini. When p53 binds to DNA, it wraps around the DNA helix with all four N-termini pointing away from the face of the core domain DNA complex.

1.5   Tetramerization and Function of p53

p53 tetramer formation is essential for its function. p53 binds DNA as a tetramer and tetramerization mediated by the C-terminus tetramerization domain allosterically regulates the DNA binding activity of p53 [15]. The DNA sequence of the p53 response element contains four pentanucleotide repeats [46]. Because each p53 DNA

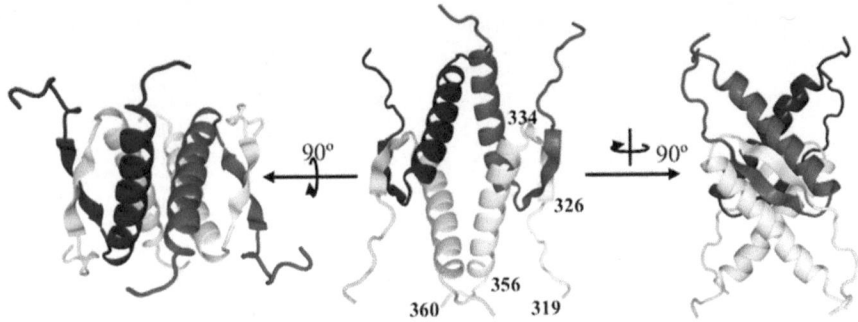

Fig. 1.5 Structure of the tetramerization domain (pdb; 3SAK). Cartoon model of p53TD (pdb; 3SAK) prepared with MolFeat version 4.0 (FiatLux Corp.). The left and right tetramers were obtained by rotating the structure in the center by 90° around the horizontal and vertical axes, respectively

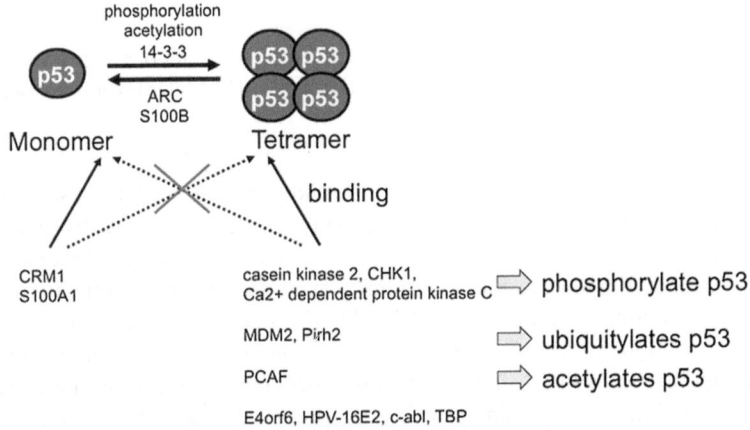

Fig. 1.6 Correlation between tetramerization and function of p53. The post-translational modifications and protein–protein interactions are dependent on the oligomeric state of p53. Conversely, post-translational modifications, such as phosphorylation and acetylation, and protein–protein interactions regulate tetramerization of p53

binding domain recognizes one repeat, a polyvalent p53 tetramer has a 100-fold higher affinity for the p53 DNA site than a p53 monomer [47]. Thus, p53 tetramerization is required for sequence-specific DNA binding. In addition, protein–protein interactions and some post-translational modifications, such as phosphorylation, acetylation, and ubiquitination, require tetramerization of p53 (Fig. 1.6). Many proteins bind directly to the tetramerization domain or have their interaction with p53 influenced by its oligomeric status. Casein kinase 2, Ca^{2+}-dependent protein kinase C and adenovirus E4orf6 directly bind to the tetramerization domain [48]. Other proteins, such as MDM2, HPV-16 E2, c-abl, and TBP, only interact with tetrameric p53. Several post-translational modifications are dependent on the

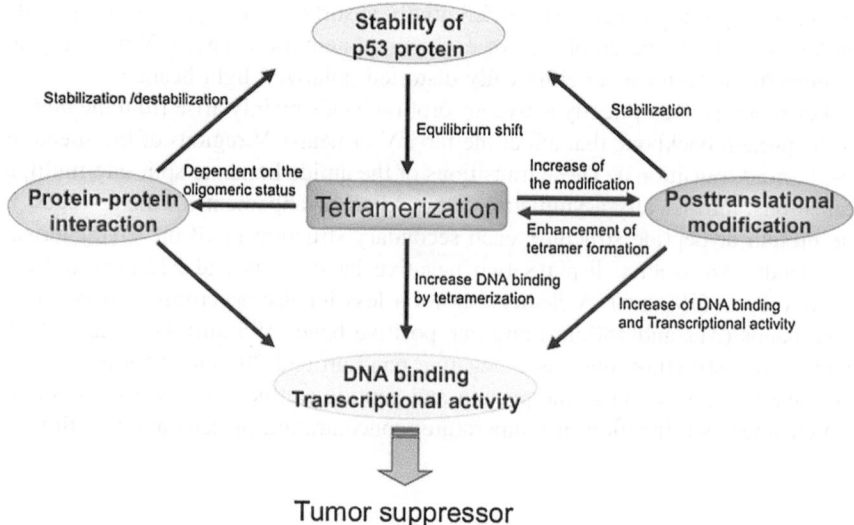

Fig. 1.7 Correlation between tetramerization and function of p53. Tetramer formation of p53 affects p53 activities, such as post-translational modifications, protein–protein interactions and DNA binding affinity. Conversely, post-translational modification and protein concentration increase or decrease p53 tetramer formation

quaternary structure of p53 [49]. The phosphorylation by CHK1 or casein kinase 2 and the ubiquitination of p53 require its tetramerization. The binding of S100 family proteins depends on the oligomeric status of p53 and controls the balance between monomer and tetramer [50]. The S100B protein interacts with the C-terminal region of p53 (residues 319–393) and protects p53 from thermal denaturation [51]. Moreover, because the nuclear export signal is located in the tetramerization domain, localization of p53 is dependent on its oligomeric status [52]. Post-translational modification affects tetramerization. The phosphorylation of Ser392 enhances tetramer formation [30, 53, 54]. Thus, tetramerization, post-translational modification, protein–protein interaction, and protein concentration regulate one another, and they control p53 tumor suppressor activity (Fig. 1.7).

1.6 Biophysical Methods

1.6.1 Circular Dichroism (CD)

CD spectroscopy measures differences in absorbance of right and left circularly polarized light. When the circularly polarized light passes through an optically active sample, the right-handed and the left-handed components interact in a different manner with the optically active molecules. Thus, the right-handed and the left-handed components are differentially absorbed. After passing through the sample, the phase and amplitude of each component are changed. When they are combined, the result is an elliptically distorted polarized light beam.

For proteins, the optically active absorption bands mainly arise from the amides of the protein backbone that affect the far-UV or near-UV regions of the spectrum (n-π*, π-π* transitions). These transitions of the amide bond are split into multiple transitions, intrinsically asymmetric or periodical arrangements of the residues. In the protein or peptide structure, each secondary structure motif has characteristic CD bands. An α-helix displays two negative bands (208 and 222 nm) and one positive band (190 nm). A β-sheet shows a less intense spectrum with two negative bands (217 and 180 nm) and one positive band (195 nm). For a disordered random coil structure, there is a negative band around 200 nm. Thermodynamic and kinetic information of the protein and peptide can be obtained by measuring CD changes as a function of temperature, concentration of denaturant or time.

1.6.2 The p53 Tetramerization Domain Peptides

The p53 tetramer structure is in equilibrium with unfolded monomer. Unfolding of p53TD has been described as a two-state transition with the tetramer (N_4; wild-type) or dimer (N_2; dimer mutants) directly converted to denatured monomers (U) [55].

$$N_4 \rightleftarrows 4U \quad \text{or} \quad N_2 \rightleftarrows 2U$$

The equilibrium constants for tetrameric or dimeric p53TD unfolding, K_u, and the free energy of unfolding, ΔG_u, are defined as:

$$K_u = [U]^4/[N_4] = 4P_t^3 f_u^4/(1 - f_u) \quad \text{Tetramer}$$
$$K_u = [U]^2/[N_2] = 2P_t f_u^2/(1 - f_u) \quad \text{Dimer}$$
$$\Delta G_u = -RT \ln K_u$$

P_t is the total p53TD monomer concentration, and f_u the fraction of denaturation.

f_u was calculated from the corresponding experimental ellipticity value θ using the expression:

$$f_u = [\theta - (\theta_{n0} + m_n T)]/[\theta_{u0} + m_u T] - [(\theta_{n0} + m_n T)]$$

θ_{n0} and θ_{u0}, the ellipticity values corresponding to the native (n) and denatured (u) states extrapolated to $T = 0$, and m_n and m_u, the slopes of the baselines preceding and following the transition region, were obtained by linear regression analysis of the baselines.

To calculate T_m and ΔH_u^{Tm}, ΔG_u values obtained for T values within the transition zone of the denaturation curve using the following equations were used to fit the appropriate expanded form of the Gibbs equation:

$$\Delta G_u = \Delta H_u^{Tm}(1 - T/T_m) + \Delta C_p[T - T_m - T\ln(T/T_m)] - RT\ln(P_t^3/2) \quad \text{Tetramer}$$
$$\Delta G_u = \Delta H_u^{Tm}(1 - T/T_m) + \Delta C_p[T - T_m - T\ln(T/T_m)] - RT\ln P_t \quad \text{Dimer}$$

where $\Delta C_p = 1.7$ kcal/(K.mol of tetramer) or 0.85 kcal/(K.mol of dimer), which correspond to 425 cal/(K.mol of monomer) [44].

1.7 Aims of this Study

p53 is a sequence-specific transcription factor that suppresses tumor development. In response to genotoxic stress, p53 induces cell cycle arrest and apoptosis by transactivating expression of downstream target genes. Furthermore, p53 also regulates microRNA processing and apoptosis in mitochondria. Tetramer formation of p53 is essential for its activity.

The aims of this study were to reveal the dysfunction threshold of p53 due to the destabilization of its tetrameric structure. The effects of tumor-associated mutations in the tetramerization domain on the tetrameric structure were analyzed. At endogenous p53 levels the mutations significantly decreased the fraction of tetramer, leading to dysfunction of p53 transcriptional activity. Based on this structural analysis, functional control of p53 via tetramer formation was also performed; (1) the tumor-associated mutants were rescued by stabilization of the tetrameric structure using calixarene derivatives and (2) transcriptional activity of p53 was inhibited by hetero-oligomerization. p53 plays important roles in cell cycle arrest, apoptosis, DNA repair, and microRNA maturation in response to genotoxic stress. Loss of p53 function, either directly through mutation or indirectly through several mechanisms, is an important step in tumorigenesis. Thus, the development of methods that enhance reduced p53 activity is important to cancer therapy. Recently, however, inactivation of p53 has also received attention because heart failure is associated with up-regulation of p53 function. In addition, temporary suppression of p53 has been suggested as an approach to reduce side effects of cancer treatment and to increase the efficiency of iPS cell generation, which is inhibited by p53–p21 pathways.

References

1. Meek DW (2009) Tumour suppression by p53: a role for the DNA damage response? Nat Rev Cancer 9:714–723
2. Meek DW, Anderson CW (2009) Posttranslational modification of p53: cooperative integrators of function. Cold Spring Harb Perspect Biol 1:a000950

3. Rodier F, Campisi J, Bhaumik D (2007) Two faces of p53: aging and tumor suppression. Nucleic Acids Res 35:7475–7484

4. Haupt Y, Maya R, Kazaz A, Oren M (1997) Mdm2 promotes the rapid degradation of p53. Nature 387:296–299

5. Picksley SM, Lane DP (1993) The p53-mdm2 autoregulatory feedback loop: a paradigm for the regulation of growth control by p53? BioEssays 15:689–690

6. Speidel D (2010) Transcription-independent p53 apoptosis: an alternative route to death. Trends Cell Biol 20:14–24

7. Suzuki HI, Yamagata K, Sugimoto K, Iwamoto T, Kato S, Miyazono K (2009) Modulation of microRNA processing by p53. Nature 460:529–533

8. Sano M, Minamino T, Toko H, Miyauchi H, Orimo M, Qin Y, Akazawa H, Tateno K, Kayama Y, Harada M, Shimizu I, Asahara T, Hamada H, Tomita S, Molkentin JD, Zou Y, Komuro I (2007) p53-induced inhibition of Hif-1 causes cardiac dysfunction during pressure overload. Nature 446:444–448

9. Komarov PG, Komarova EA, Kondratov RV, Christov-Tselkov K, Coon JS, Chernov MV, Gudkov AV (1999) A chemical inhibitor of p53 that protects mice from the side effects of cancer therapy. Science 285:1733–1737

10. Hong H, Takahashi K, Ichisaka T, Aoi T, Kanagawa O, Nakagawa M, Okita K, Yamanaka S (2009) Suppression of induced pluripotent stem cell generation by the p53-p21 pathway. Nature 460:1132–1135

11. Zhang Y, Xiong Y (2001) A p53 amino-terminal nuclear export signal inhibited by DNA damage-induced phosphorylation. Science 292:1910–1915

12. Anderson ME, Woelker B, Reed M, Wang P, Tegtmeyer P (1997) Reciprocal interference between the sequence-specific core and nonspecific C-terminal DNA binding domains of p53: implications for regulation. Mol Cell Biol 17:6255–6264

13. Mazur SJ, Sakaguchi K, Appella E, Wang XW, Harris CC, Bohr VA (1999) Preferential binding of tumor suppressor p53 to positively or negatively supercoiled DNA involves the C-terminal domain. J Mol Biol 292:241–249

14. Hupp TR, Meek DW, Midgley CA, Lane DP (1992) Regulation of the specific DNA binding function of p53. Cell 71:875–886

15. Balagurumoorthy P, Sakamoto H, Lewis MS, Zambrano N, Clore GM, Gronenborn AM, Appella E, Harrington RE (1995) Four p53 DNA-binding domain peptides bind natural p53-response elements and bend the DNA. Proc Natl Acad Sci U S A 92:8591–8595

16. Halazonetis TD, Kandil AN (1993) Conformational shifts propagate from the oligomerization domain of p53 to its tetrameric DNA binding domain and restore DNA binding to select p53 mutants. EMBO J 12:5057–5064

17. Nagaich AK, Zhurkin VB, Durell SR, Jernigan RL, Appella E, Harrington RE (1999) p53-induced DNA bending and twisting: p53 tetramer binds on the outer side of a DNA loop and increases DNA twisting. Proc Natl Acad Sci U S A 96:1875–1880

18. Anderson CW, Appella E, Sakaguchi K (1998) Posttranslational modifications involved in the DNA damage response. J Protein Chem 17:527

19. Bulavin DV, Saito S, Hollander MC, Sakaguchi K, Anderson CW, Appella E, Fornace AJJ (1999) Phosphorylation of human p53 by p38 kinase coordinates N-terminal phosphorylation and apoptosis in response to UV radiation. EMBO J 18:6845–6854

20. Canman CE, Lim DS, Cimprich KA, Taya Y, Tamai K, Sakaguchi K, Appella E, Kastan MB, Siliciano JD (1998) Activation of the ATM kinase by ionizing radiation and phosphorylation of p53. Science 281:1677–1679

21. Higashimoto Y, Saito S, Tong XH, Hong A, Sakaguchi K, Appella E, Anderson CW (2000) Human p53 is phosphorylated on serines 6 and 9 in response to DNA damage-inducing agents. J Biol Chem 275:23199–23203

22. Lees-Miller SP, Sakaguchi K, Ullrich SJ, Appella E, Anderson CW (1992) Human DNA-activated protein kinase phosphorylates serines 15 and 37 in the amino-terminal transactivation domain of human p53. Mol Cell Biol 12:5041–5049

23. Siliciano JD, Canman CE, Taya Y, Sakaguchi K, Appella E, Kastan MB (1997) DNA damage induces phosphorylation of the amino terminus of p53. Genes Dev 11:3471–3481
24. Sakaguchi K, Saito S, Higashimoto Y, Roy S, Anderson CW, Appella E (2000) Damage-mediated phosphorylation of human p53 threonine 18 through a cascade mediated by a casein 1-like kinase. Effect on Mdm2 binding. J Biol Chem 275:9278–9283
25. Kar S, Sakaguchi K, Shimohigashi Y, Samaddar S, Banerjee R, Basu G, Swaminathan V, Kundu TK, Roy S (2002) Effect of phosphorylation on the structure and fold of transactivation domain of p53. J Biol Chem 277:15579–15585
26. Polley S, Guha S, Roy NS, Kar S, Sakaguchi K, Chuman Y, Swaminathan V, Kundu T, Roy S (2008) Differential recognition of phosphorylated transactivation domains of p53 by different p300 domains. J Mol Biol 376:8–12
27. Pise-Masison CA, Radonovich M, Sakaguchi K, Appella E, Brady JN (1998) Phosphorylation of p53: a novel pathway for p53 inactivation in human T-cell lymphotropic virus type 1-transformed cells. J Virol 72:6348–6355
28. Anderson CW, Appella E (2011) Signaling to the p53 tumor suppressor through pathways activated by genotoxic and nongenotoxic stress. In: Bradshaw RA, Dennis EA (eds) Regulation in Organella and Cell Compartment Signaling, Chap 264, Elsevier BV, CA, pp 235–254
29. Sakaguchi K, Herrera JE, Saito S, Miki T, Bustin M, Vassilev A, Anderson CW, Appella E (1998) DNA damage activates p53 through a phosphorylation-acetylation cascade. Genes Dev 12:2831–2841
30. Sakaguchi K, Sakamoto H, Lewis MS, Anderson CW, Erickson JW, Appella E, Xie D (1997) Phosphorylation of serine 392 stabilizes the tetramer formation of tumor suppressor protein p53. Biochemistry 36:10117–10124
31. Li AG, Piluso LG, Cai X, Wei G, Sellers WR, Liu X (2006) Mechanistic insights into maintenance of high p53 acetylation by PTEN. Mol Cell 23:575–587
32. Pavithra L, Mukherjee S, Sreenath K, Kar S, Sakaguchi K, Roy S, Chattopadhyay S (2009) SMAR1 forms a ternary complex with p53-MDM2 and negatively regulates p53-mediated transcription. J Mol Biol 388:691–702
33. Kussie PH, Gorina S, Marechal V, Elenbaas B, Moreau J, Levine AJ, Pavletich NP (1996) Structure of the MDM2 oncoprotein bound to the p53 tumor suppressor transactivation domain. Science 274:948–953
34. Di Lello P, Jenkins LM, Jones TN, Nguyen BD, Hara T, Yamaguchi H, Dikeakos JD, Appella E, Legault P, Omichinski JG (2006) Structure of the Tfb1/p53 complex: Insights into the interaction between the p62/Tfb1 subunit of TFIIH and the activation domain of p53. Mol Cell 22:731–740
35. Rustandi RR, Baldisseri DM, Weber DJ (2000) Structure of the negative regulatory domain of p53 bound to S100B(betabeta). Nat Struct Biol 7:570–574
36. Joerger AC, Fersht AR (2008) Structural biology of the tumor suppressor p53. Annu Rev Biochem 77:557–582
37. Sakamoto H, Lewis MS, Kodama H, Appella E, Sakaguchi K (1994) Specific sequences from the carboxyl terminus of human p53 gene product form anti-parallel tetramers in solution. Proc Natl Acad Sci U S A 91:8974–8978
38. Clore GM, Omichinski JG, Sakaguchi K, Zambrano N, Sakamoto H, Appella E, Gronenborn AM (1994) High-resolution structure of the oligomerization domain of p53 by multidimensional NMR. Science 265:386–391
39. Clubb RT, Omichinski JG, Sakaguchi K, Appella E, Gronenborn AM, Clore GM (1995) Backbone dynamics of the oligomerization domain of p53 determined from 15 N NMR relaxation measurements. Protein Sci 4:855–862
40. Miller M, Lubkowski J, Rao JKM, Danishefsky AT, Omichinski JG, Sakaguchi K, Sakamoto H, Appella E, Gronenborn AM, Clore GM (1996) The oligomerization domain of p53: crystal structure of the trigonal form. FEBS Lett 399:166–170

41. Clore GM, Ernst J, Clubb R, Omichinski JG, Kennedy WMP, Sakaguchi K, Appella E, Gronenborn AM (1995) Refined solution structure of the oligomerization domain of the tumour suppressor p53. Nat Struct Biol 2:321–333
42. Clore GM, Omichinski JG, Sakaguchi K, Zambrano N, Sakamoto H, Appella E, Gronenborn AM (1995) Interhelical angles in the solution structure of the oligomerization domain of p53: correction. Science 267:1515–1516
43. Jeffrey PD, Gorina S, Pavletich NP (1995) Crystal structure of the tetramerization domain of the p53 tumor suppressor at 1.7 angstroms. Science 267:1498–1502
44. Johnson CR, Morin PE, Arrowsmith CH, Freire E (1995) Thermodynamic analysis of the structural stability of the tetrameric oligomerization domain of p53 tumor suppressor. Biochemistry 34:5309–5316
45. Tidow H, Melero R, Mylonas E, Freund S, Grossmann JG, Carazo JM, Svergun DI, Valle M, Fersht AR (2007) Quaternary structures of tumor suppressor p53 and a specific p53–DNA complex. Proc Natl Acad Sci U S A 104:12324–12329
46. Malecka KA, Ho WC, Marmorstein R (2009) Crystal structure of a p53 core tetramer bound to DNA. Oncogene 28:325–333
47. McLure KG, Lee PW (1998) How p53 binds DNA as a tetramer. EMBO J 17:3342–3350
48. Chene P (2001) The role of tetramerization in p53 function. Oncogene 20:2611–2617
49. Ullrich SJ, Sakaguchi K, Lees-Miller SP, Fiscella M, Mercer WE, Anderson CW, Appella E (1993) Phosphorylation at Ser-15 and Ser-392 in mutant p53 molecules from human tumors is altered compared to wild-type p53. Proc Natl Acad Sci U S A 90:5954–5958
50. van Dieck J, Fernandez–Fernandez MR, Veprintsev DB, Fersht AR (2009) Modulation of the oligomerization state of p53 by differential binding of proteins of the S100 family to p53 monomers and tetramers. J Biol Chem 284:13804–13811
51. Delphin C, Ronjat M, Deloulme JC, Garin G, Debussche L, Higashimoto Y, Sakaguchi K, Baudier J (1999) Calcium-dependent interaction of S100B with the C-terminal domain of the tumor suppressor p53. J Biol Chem 274:10539–10544
52. Stommel JM, Marchenko ND, Jimenez GS, Moll UM, Hope TJ, Wahl GM (1999) A leucine-rich nuclear export signal in the p53 tetramerization domain: regulation of subcellular localization and p53 activity by NES masking. EMBO J 18:1660–1672
53. Sakaguchi K, Sakamoto H, Xie D, Erickson JW, Lewis MS, Anderson CW, Appella E (1997) Effect of phosphorylation on tetramerization of the tumor suppressor protein p53. J Protein Chem 16:553–556
54. Sakamoto H, Kodama H, Higashimoto Y, Kondo M, Lewis MS, Anderson CW, Appella E, Sakaguchi K (1996) Chemical synthesis of phosphorylated peptides of the carboxy-terminal domain of human p53 by a segment condensation method. Int J Pept Protein Res 48:429–442
55. Mateu MG, Fersht AR (1998) Nine hydrophobic side chains are key determinants of the thermodynamic stability and oligomerization status of tumour suppressor p53 tetramerization domain. EMBO J 17:2748–2758

Chapter 2
Quantitative Analysis for p53 Tetramerization Domain Mutants Reveals a Low Threshold for Tumor Suppressor Inactivation

2.1 Introduction

Genome instability and DNA breakage are the hallmarks of cancer cells that arise in response to the activation of oncogenes through point mutations, gene amplifications, or gene translocations [1, 2]. Counterbalancing the effects of oncoproteins are tumor suppressor proteins, the most important of which is p53, a transcription factor that modulates cell cycle arrest, senescence, apoptosis, and DNA repair largely via the direct or indirect induction or repression of hundreds of genes [3].

The p53 tumor suppressor monomer is a 393 amino acid protein with five domains: An N-terminal transactivation domain (91–42); a proline- rich domain (61–92); a central site-specific DNA-binding domain (101–300); a tetramerization domain (TD, 326–356); and a C-terminal basic domain (364–393). Several stressors, including DNA damage, activate p53 partly through multiple post-translational modifications modulating its activity and stability [4]. However, wild-type p53 acts as a transcription factor only when it binds site-specific DNA response elements as a tetramer [5]. Furthermore, a number of the post-translational modifications that are believed to be important regulators of p53 activity depend on its quaternary structure [6–11]. The p53 protein also exhibits transcription-independent apoptogenesis, possibly contributing to its role in tumor suppression, that is mediated through its interaction with BCL2 family members, including Bak. The efficient targeting to and oligomerization of Bak in the mitochondrial membrane reportedly depends on p53 oligomerization [12]. Thus, tetramer formation by p53 is crucial to its tumor suppressive activity.

About half of human tumors carry inactivating mutations in the *TP53* gene [13, 14]. Unlike other tumor suppressor genes, such as *RB1*, *APC*, *BRCA1*, and *CDKN2A* that are inactivated primarily by deletion or nonsense mutations, 74% of *TP53* tumor-derived mutations are point mutations that change a single amino acid. More than 95 % of these missense mutations occur in the DNA-binding

R. Kamada, *Tetramer Stability and Functional Regulation of Tumor Suppressor Protein p53*, Springer Theses, DOI: 10.1007/978-4-431-54135-6_2, © Springer Japan 2012

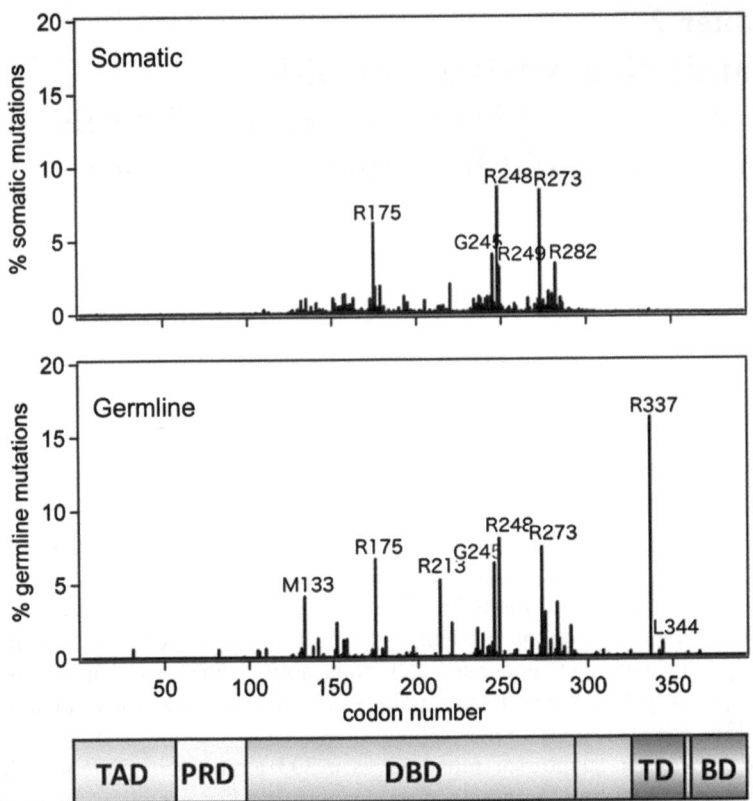

Fig. 2.1 Relative frequency of somatic (*top*) and germline (*bottom*) mutations along p53 sequence. Codon 337 in the tetramerization domain is the most frequently affected position in p53 germline mutations. From the IARC TP53 Mutation Database, release R14, November 2009)

domain; they fall into two main categories, commonly termed DNA contact- and conformational-mutations (Fig. 2.1). In contrast, about 17 % of germ-line p53 mutations in people with Li-Fraumeni syndrome and Li-Fraumeni-like ones affect amino acids in the TD even though it consists only of a short amino acid segment (≈ 30 a.a.), while ~ 80 % of germ-line mutations affect DNA-binding domain residues, viz., six times as long as the TD [14]. This finding implies that germ-line mutations exist at similar frequencies in the tetramerization and DNA-binding domains, and both are essential for p53-mediated tumor suppressor activity.

The p53TD consists of a β-strand (Glu326-Arg333), a tight turn (Gly334), and an α-helix (Arg335-Gly356) [15]. The structure of the TD was determined by NMR spectroscopy [16] and X-ray crystallography [17]. Two monomers form a dimer through their antiparallel β-sheets and α-helices, and two dimers become a tetramer through the formation of an unusual four-helix bundle. Alanine (Ala)-scanning of p53TD revealed that nine hydrophobic residues constitute critical determinants of its stability and oligomerization status [18]. An earlier study of tumor-derived mutants R337C, R337H, and L344P from patients with Li-Frau-meni-like syndrome revealed a propensity for dramatic destabilization; the presence of the R337H mutation entailed pH-dependent instability of the mutant p53 tetramer [19]. Leu344 occurs in the α-helix, and after introducing a helix-breaking proline (L344P), p53 could not form tetramers. R337C forms dimers and tetramers at low temperature; however, even though its tetrameric structure is destabilized significantly at physiological temperatures, it is only partially inactivated in several functional assays [20, 21]. The p53 proteins with these mutations, as with other p53TD mutations (e.g., L330H, R337L, R342P, E349D and G334V), exhibit an overall decrease in DNA-binding and transactivation activity [22, 23].

Because the p53 tetramer is in equilibrium with the monomer, the protein concentration of p53 will affect its oligomeric status [18, 24]. In unstressed normal cells, p53 is maintained at low levels by continuous ubiquitylation and subsequent degradation by the 26S proteasome [25]. DNA damage-induced phosphorylation of N-terminal residues of p53, and of Mdm2, an ubiquitin protein ligase, inhibits its binding to the latter and enables p53 stabilization and accumulation [4]. A high concentration of p53 shifts the monomer-tetramer equilibrium toward the tetramer state, thereby promoting increased DNA binding and interactions with proteins important for p53 activation and function, and heightening post-translational modifications that activate p53. Past research used only semi-quantitative analyses to assess the effects of mutations on the oligomeric structure and transcriptional activity of p53 [26–28]. Whilst this research determined the oligomeric status of the mutant p53 protein by cross-linking [27] or by Fluorescence Intensity Distribution Analysis (FIDA) [28], the abundance of the p53 protein was not controlled; thus, a destabilized mutant might show wild-type stability under high concentrations of mutant p53.

In this study, I quantitatively analyzed the oligomeric structure and stability of TD peptides from the reported cancer-associated, TD mutants of p53. Surprisingly, the abilities of these mutants to form tetramers spanned a broad, almost continuous distribution. While mutants that changed the domain core drastically prevented tetramer formation and/or folding as previously reported, the effects of many mutants were much more subtle. Nevertheless, even for mutants that slightly destabilized tetramer formation, at an endogenous concentration of p53, the fraction of tetramer is significantly decreased. The data further suggested that additional studies of the biochemical and biophysical properties of the TD might be required to explain why some p53 TD mutations are cancer-associated.

2.2 Experimental Procedures

2.2.1 Peptide Synthesis and Purification

WT- and mutant-p53TD peptides, comprising residues 319–358 of the extended TD, were synthesized as described previously [29]. Peptide concentrations were measured spectrophotometrically using an extinction coefficient for mutant p53TD peptides, $\varepsilon_{280} = 1280$ M^{-1} cm^{-1}, corresponding to a single tyrosine; for G334W and G356W, $\varepsilon_{280} = 6800$ M^{-1} cm^{-1}, corresponding to a single tyrosine and a tryptophan. Because the peptides Y327D, Y327H, and Y327S have no Tyr or Trp, peptide concentrations were determined by the BCA method (Pierce Co.) using a WT peptide as the standard.

2.2.2 Gel Filtration Chromatography

The WT- and mutant-p53TD peptides were resolved using a Superdex 75 PC 3.2/ 30 (GE Healthcare) with a Precision Column Holder (GE Health) in 50 mM phosphate buffer pH 7.5, 100 mM NaCl [29]. Peptide concentrations were 100 μM. The flow-rate was 0.1 mL/min at 15 °C, and the effluent at 214 nm was monitored. Each peak was quantified by calculating the peak area using IGOR software (Wavemetrics).

2.2.3 Thermal Denaturation by Circular Dichroism Spectroscopy

For the CD measurements, a Jasco-805 spectropolarimeter was employed using a 1 mm path-length quartz cell. CD spectra were recorded in 50 mM sodium phosphate buffer containing 100 mM NaCl, pH 7.5. For the thermal denaturation studies, spectra were recorded at discrete temperatures from 4 to 96 °C with a scan rate of 1 °C/min; ellipticity was measured at 222 nm for the p53TD solutions (10 μM monomer in 50 mM phosphate buffer, pH 7.5). The unfolding process of the p53TD peptide was fitted to a two-state transition model wherein the native tetramer directly converts to an unfolded monomer, as previously described [18, 24]. The thermodynamic parameters of the peptides were determined by calculation with the functions described by Mateu et al. [18]. The T_m and ΔH^{Tm} was determined by fitting the fraction of monomer; we estimated the K_d value of the tetramer-monomer transition from $K_d = ((1-K_u)/2)^{-1/3}$ [30]. For dimer mutants, $K_d = K_u^{-1}$ was used. The oligomeric states at 37 °C against the peptide concentration were assessed via the K_d value.

2.2.4 Structural Modeling of p53TD Mutants

The three-dimensional coordinates of p53TD wild-type (PDB: 3sak) were used as a template. Homology modeling of mutants was performed with Modeller software [31].

Table 2.1 Missense mutations found in tetramerization domain

mutation			tumor type		mutation			tumor type
Glu	326	Gly	Skin	β-strand	Arg	337	Pro	Breast, Lung, Ovary
Tyr	327	Asp	Colorectum		Phe	338	Ile	Breast
		His	Skin				Leu	Adrenal Gland
		Ser	Breast		Glu	339	Gln	Breast
Phe	328	Leu	Skin				Lys	Head and Neck
		Ser	Colorectum		Phe	341	Cys	Breast, Prostate Gland
		Val	Stomach		Arg	342	Gln	Larynx, Prostate Gland
Thr	329	Ile	Head and Neck				Leu	Brain
		Ser	Ovary				Pro	Breast, Ovary
Leu	330	Arg	Breast		Glu	343	Gly	Colon, Lung
		His	Liver, Ovary		Leu	344	Arg	Bladder
		Pro	Colorectum				Pro	Hypopharynx
Gln	331	Arg	Prostate Gland		Glu	346	Ala	Nasopharynx
		His	Breast, Lung		Ala	347	Gly	Brain
		Pro	Esophagus				Thr	Lung, Breast, Ovary
Ile	332	Val	Ovary		Leu	348	Phe	Lung
Gly	334	Trp	Lung				Ser	Oropharynx
		Val	Lung	turn	Glu	349	Asp	Bladder
Arg	335	Gly	Pancreas	α-helix	Lys	351	Asn	Ovary
		His	Lung		Asp	352	His	Esophagus
		Leu	Other and Unspecified Parts of Mouth		Ala	353	Thr	Bladder
Arg	337	Cys	Bones, Brain, Breast, Colorectum, Esophagus		Gln	354	Arg	Kidney
							Glu	Lymph Nodes
		His	Adrenal Gland, Liver				Lys	Head and Neck
		Leu	Breast, Prostate Gland		Gly	356	Ala	Pancreas
							Trp	Corpus Uteri, Vulva

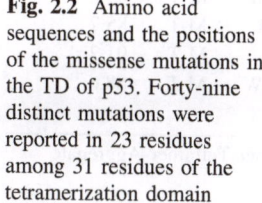
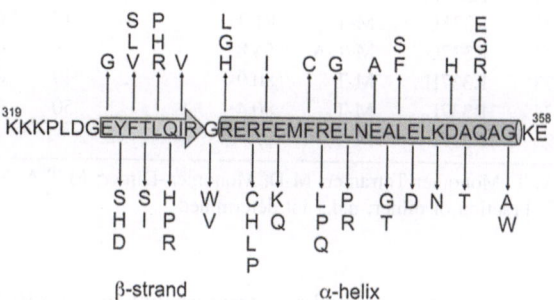

Fig. 2.2 Amino acid sequences and the positions of the missense mutations in the TD of p53. Forty-nine distinct mutations were reported in 23 residues among 31 residues of the tetramerization domain

2.3 Results

2.3.1 Oligomerization State of Mutant p53 Tetramerization Domains

Fifty distinct mutations in human cancers occur in 25 of the 31 residues that comprise the p53 core TD (amino acids 326–356) (Table 2.1, Figs. 2.2, 2.3). Wild-type (WT) and mutant p53TD peptides corresponding to residues 319–358 were synthesized and their oligomeric state and thermodynamic stability were analyzed; I quantified this state from the peak areas corresponding to a monomer and tetramer during gel-filtration chromatography (Table 2.2). WT and most

Table 2.2 The fraction of tetramer determined by gel filtration chromatography

No.	mutant	State	Tetramer (%)	No.	mutant	State	Tetramer (%)
	WT	M-T	90.7	26	F338I	M-T	65.9
1	E326G	M-T	87.8	27	F338L	M-T	75.2
2	Y327D	M-T	85.0	28	E339 K	M-T	83.2
3	Y327H	M-T	87.1	29	E339Q	M-T	84.0
4	Y327S	M-T	81.1	30	F341C	M-D	55.4*
5	F328L	M-T	88.2	31	R342L	M-T	90.4
6	F328S	M-T	74.4	32	R342P	M	0.0
7	F328 V	M-T	83.5	33	R342Q	M-T	88.8
8	T329I	M-T	84.5	34	E343G	M-T	78.1
9	T329S	M-T	89.7	35	L344P	M	0.0
10	L330H	M-T	45.1	36	L344R	M-D	61.0*
11	L330P	M	0.0	37	E346A	M-T	81.4
12	L330R	M	0.0	38	A347G	M-T	88.3
13	Q331H	M-T	85.6	39	A347T	M-D	62.1[a]
14	Q331P	M-T	88.4	40	L348F	M-T	76.3
15	Q331R	M-T	90.9	41	L348S	M-T	40.5
16	I332 V	M-T	88.5	42	E349D	M-T	89.5
17	G334 V	M-T-A	84.3	43	K351 N	n.d.	n.d.
18	G334 W	M-T	82.0	44	D352H	M-T	89.7
19	R335G	M-T	62.2	45	A353T	M-T	76.9
20	R335H	M-T	89.5	46	Q354E	M-T	85.2
21	R335L	M-T	81.3	47	Q354 K	M-T	82.2
22	R337C	M-T-A	54.8	48	Q354R	M-T	85.7
23	R337H	M-T	80.9	49	G356A	M-T	91.2
24	R337L	M-T	80.4	50	G356 W	M-T	88.6
25	R337P	M	0.0				

M-T, Monomer-Tetramer; M-D, Monomer–Dimer; M-T-A, Monomer-Tetramer-Aggregate
[a] Fraction of dimer; n.d., not determined

mutant peptides eluted as tetramers, but five, L330P, L330R, R337P, L342P, and L344P, eluted as a single peak contemporaneously with the monomer mutant L330A. Interestingly, three mutants (F341C, L344R, and A347T) eluted between the tetramer and monomer fractions. Accordingly, five mutants, L330R/P, R337P, L342P, and L344P, exist as monomers, three mutants, F341C, L344R, and A347T as dimers, and the others as tetramers under conditions used in this study. Moreover, some mutants, such as L330H, R337C, and L348S, contained a lower tetramer fraction (45.1, 54.8, and 40.5 %, respectively), and part of these peptides chromatographed as monomers, implying destabilization of their tetrameric structure, thus favoring the monomer side of the monomer-tetramer equilibrium.

2.3.2 Secondary Structure of Mutant p53 Tetramerization Domains

The secondary structures of all mutant peptides were deduced from their CD spectra (Fig. 2.3). Five monomeric mutants (L330P, L330R, R337P, R342P, and L344P) showed a negative minimum near 200 nm, characteristic of a random coil, even under a high (10 μM) peptide concentration and low temperature of 4 °C. Seemingly, substitutions by Pro catastrophically affect tetramer formation. Five mutants (L330H, Q331P, R337C, F338I, and L348S) showed weaker negative CD spectra between ∼210 and 240 nm compared with the WT, pointing to destabilized WT-like tetrameric structures. The three dimer mutants (F341C, L344R, and A347T) displayed the same spectra as the other tetrameric mutants, indicating that their α-helical segment and β-strand are structurally similar to those of the WT tetramer. Still other mutant peptides formed WT-like tetrameric structures under these same conditions.

2.3.3 Thermal Stability of the Mutant p53 Tetramerization Domains

The effects of temperature on the conformation of the WT and mutant peptides were analyzed by calculating the thermal denaturation curves for each p53 peptide from changes in CD ellipticity at 222 nm, using a two-state transition mode. Figures 2.4, 2.5 and Table 2.3 show that the effects of amino acid changes on the tetramer's stability elicited alterations in the ΔT_ms of the mutant peptides from 4.8 to −46.8 °C. Changes to the TD hydrophobic core residues (F328, L330, R337, F338, F341, L344, and L348), except for I332V, dramatically lowered stability; modifications to solvent-exposed residues had less profound effects. The introduction of proline into the α-helix (R337P, R342P, and L344P) devastated tetramer formation; these peptides existed substantially as monomers only. No cancer-associated mutation has been reported in the codon for the hydrophobic core residue Met340. Four cancer-associated mutants changed amino acids such as to actually increase tetramer stability (T329I, Q331H, Q331R, and G356A; Table 2.3, Fig. 2.4). I noted a good correlation between the fraction of oligomers analyzed by gel filtration and the T_m value of the mutants obtained by CD (Fig. 2.11), indicating that these thermodynamic parameters corresponded to the tetrameric state of the p53 peptides.

2.3.4 Effects of Mutation on the Tetrameric Structure

Especially, Pro mutations in the α-helix caused devastating effects on tetramer formation and these peptides existed only as the monomer (Fig. 2.3). Mutations of

Fig. 2.3 Space-filling model of p53TD (pdb; 3sak) prepared with MolFeat version 4.0 (FiatLux Corp.) The amino acid residues of the mutation site in the p53TD and the location of these residues in the tetrameric structure are shown. The primary dimers are depicted, and the other dimer is removed to give a direct view of the protein's interior. The bottom dimer was obtained by rotating the structure in the top picture by 180° around the vertical axis

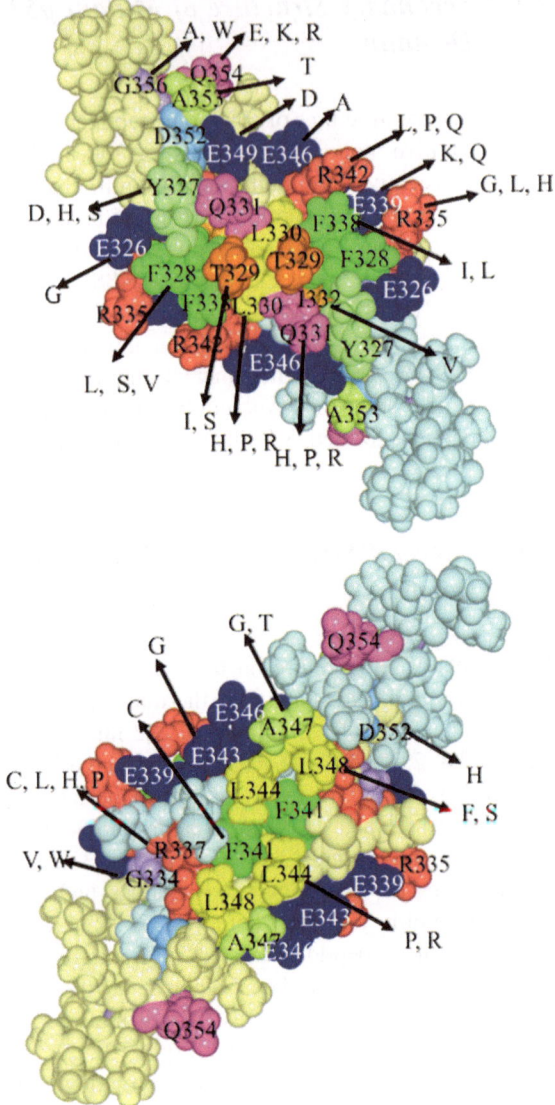

amino acid residues in the hydrophobic core of the p53 tetramerization domain induced dramatic reduction in stability of the structures. On the other hand, mutations of the residues accessible to solvent were less effective in destabilizing the tetrameric structures. Interestingly, the stability of the mutants was highly dispersed and there was no large stability gap between the mutants. These results suggested that similar to the case of the mutations (R337H, L344P) found in Li-Fraumeni syndrome, other mutations cause destabilization of the tetrameric structure and are associated with dysfunction of the tumor suppressor activity of

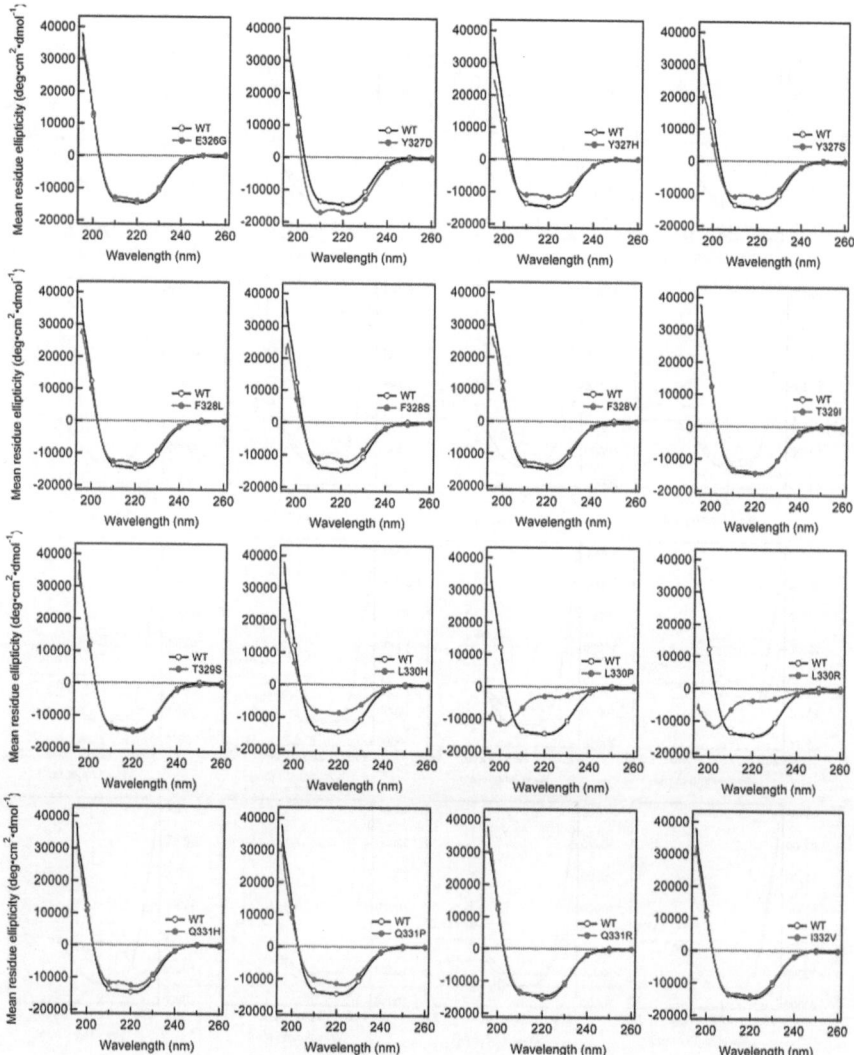

Fig. 2.4 CD spectra of WT and mutant p53TD peptides. CD spectra of WT and mutant p53TD peptides in 50 mM phosphate buffer, pH 7.5, 100 mM NaCl at 4 °C. Peptide concentration was 10 μM

p53 protein. Thus, the threshold resulting in destabilization of the structure and hence loss of p53 tumor suppressor function could be extremely low.

Three mutants (F341C, L344R, and A347T), which formed dimers, were strongly destabilized (T_m = 23.8–44.3 °C) (Fig. 2.6). F341C was the most destabilized mutant out of the three (T_m = 23.8 °C). Phe341 is located in the hydrophobic core and the hydrophobicity of the Phe residue stabilized the tetrameric structure of p53. The Phe341 residue from one peptide chain is located near another Phe341 residue

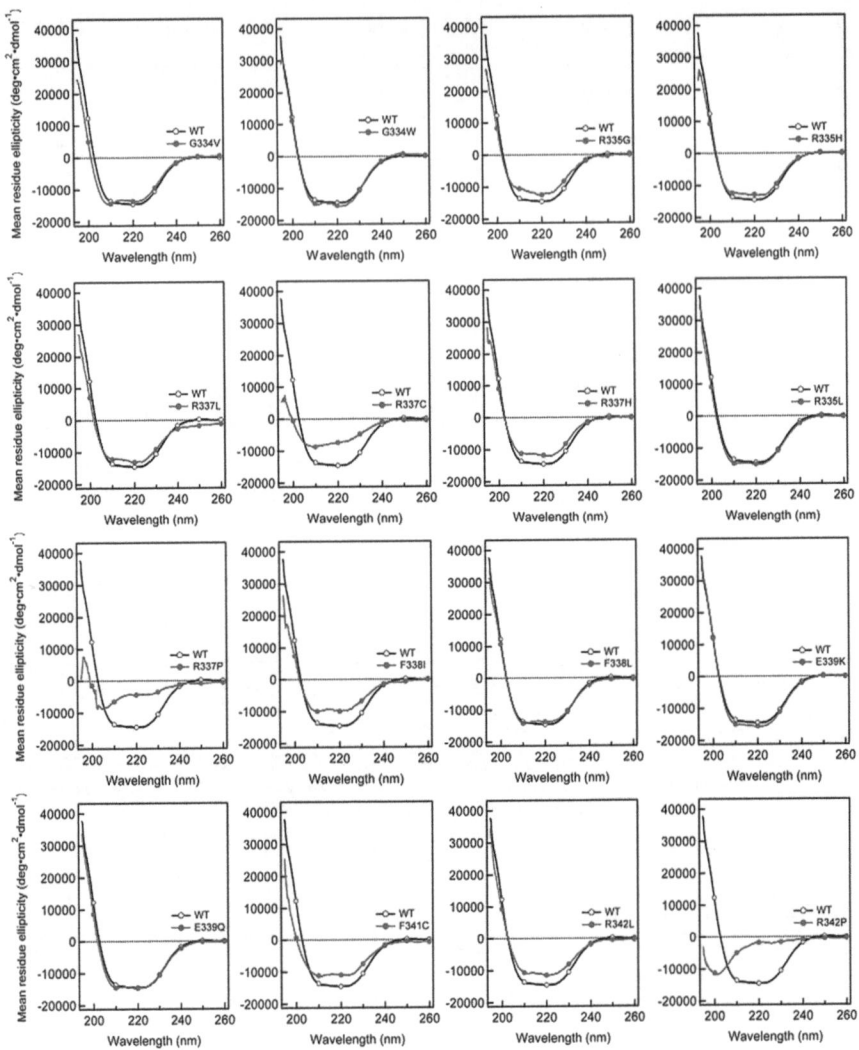

Fig. 2.4 continued

from the opposite side-chain. The mutation of Phe341 to Cys might result in the ability of C341 to form a disulfide bond in the tetrameric structure. Leu344 and Ala347 are located at the dimer–dimer interface, and Leu344 also constitutes part of the hydrophobic core of the tetramer. It is clear that mutation of Leu344 to Arg resulting in a charged polar side chain disrupts the tetrameric structure of p53. The Ala347 residue from one peptide chain and a residue from another peptide chain are in an opposite position. A347G could form tetramers although its stability was moderately destabilized. In contrast, A347T could not form tetramers because of the substitution with Thr, which has a more bulky side chain than Gly.

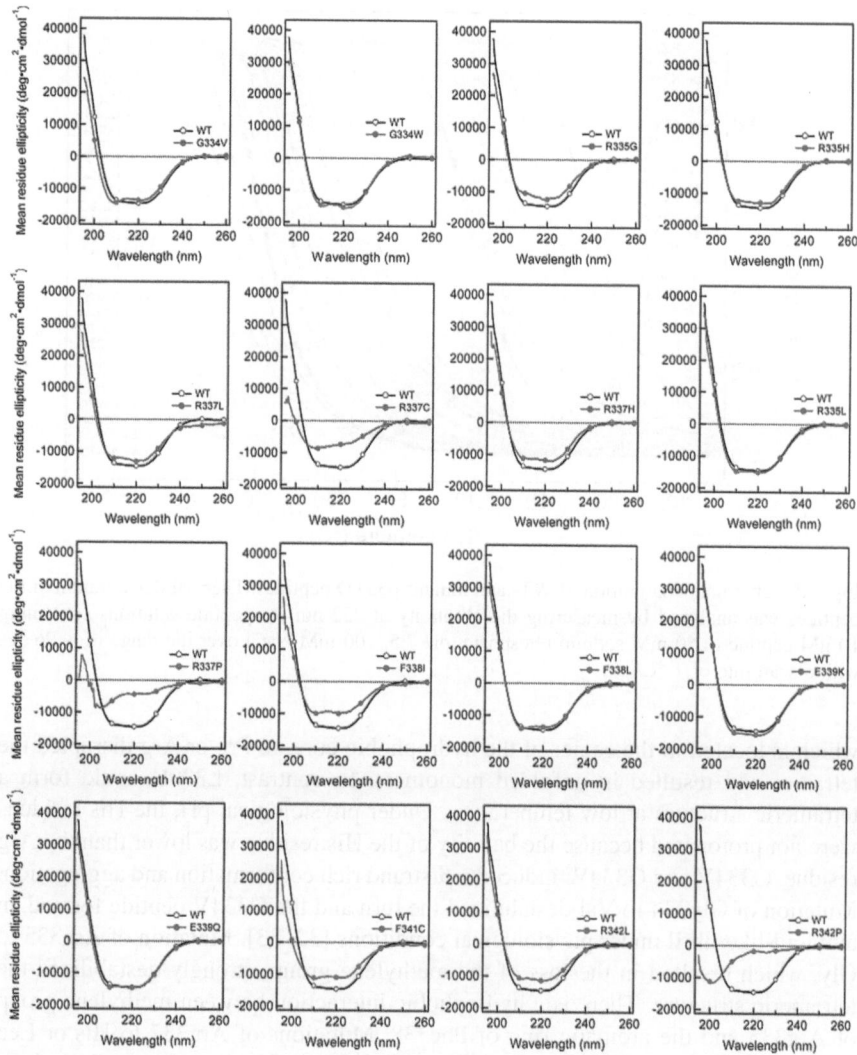

Fig. 2.4 continued

Ten mutant peptides (F328V, F328S, L330H, G334V, R335G, R337C, R337L, R337H, F338I, and L348S) formed tetramers at 10 µM concentration of the monomer and at 4 °C, however, the stability of the tetrameric structures were significantly low (T_m = 21.6–49.9 °C) (Fig. 2.7). Phe328 and Phe338 are located in the hydrophobic core and these amino acid residues stabilized the tetrameric structure through a π-interaction. Mutation of Phe328 to Val or Phe338 to Ile strongly destabilized the structure (T_m = 37.9 and 36.8 °C, respectively) even if Val and Ile were hydrophobic amino acids. This suggests that the aromaticity of Phe328 and Phe338 is important for the tetrameric structure. Mutation of Leu330,

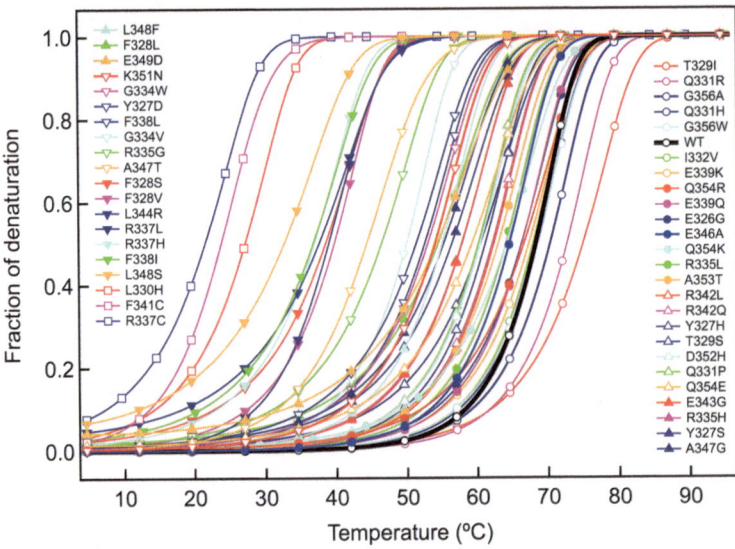

Fig. 2.5 Thermal denaturation of WT- and mutant-p53TD peptides. Thermal denaturation of the peptides was analyzed by measuring the ellipticity at 222 nm for peptide solutions containing 10 μM peptide in 50 mM sodium phosphate, pH 7.5, 100 mM NaCl over the range of 4–96 °C, with a scan rate of 1 °C/min

which is located in the center of the hydrophobic core, to Pro or Arg disrupted the tetramer and resulted in unfolded monomers. In contrast, L330H could form a tetrameric structure at low temperature. Under physiological pH, the His residues were not protonated because the basicity of the His residue was lower than the Arg residue. G334V and G334W induced a β-strand rich conformation and aggregation. Mutation of Gly334 to Val destabilized the turn and the G334V peptide formed an amyloid-like fibril under physiological conditions [32, 33]. Mutation of Arg335 to Gly, which resulted in the loss of the methylene group, strongly destabilized the tetrameric structure. There is a hydrophobic interaction between methylene group of Arg335 and the aromatic ring of Phe338. Mutations of Arg337 to His or Leu resulted in a strong destabilization (T_m = 36.9 and 37.6 °C, respectively) of the structure. Arg337 forms a salt-bridge with Asp352, and the methylene group of Arg337 likely interacts with nearby hydrophobic residues. Mutation of Arg337 to Leu was more destabilizing than mutation to His. The stability of R337H has been shown to be highly sensitive to pH within the physiological range [33]. Under the conditions used in the CD experiments, His had an overall positive charge like Arg. Therefore, R337H was slightly more stable than R337L. Mutation of the Leu348 residue which is buried at the tetramer interface, to Ala resulted in a dimer not a tetramer [18, 34]. This suggested that Leu348 stabilized the tetrameric structure through hydrophobic interactions. Mutation of Leu348 to Ser induced significant destabilization of the structure. In contrast, mutation to a hydrophobic amino acid residue Phe had little impact on the structure.

Table 2.3 Thermodynamic parameters for the mutant peptides

No.	mutant	State	T_m (°C)	ΔH_u^{Tm} (kcal/mol)	$\Delta\Delta G_u^{Tm}$ (kcal/mol)	K_d (nM)	WT AA
1	WT	M-T	68.4 ± 0.3	166.0 ± 7.0	0.0	10.2	solvent-exposed
2	E326G	M-T	66.3 ± 0.2	134.0 ± 3.8	0.8	63.5	intermonomer with 331, 333
3	Y327D	M-T	52.6 ± 0.1	111.8 ± 2.6	6.1	787.3	intermonomer with 331, 333
4	Y327H	M-T	61.2 ± 0.3	135.7 ± 6.3	3.0	113.1	intermonomer with 331, 333
5	Y327S	M-T	56.4 ± 0.1	102.6 ± 2.1	4.1	647.2	intermonomer with 331, 333
6	F328L	M-T	54.5 ± 0.3	94.1 ± 3.7	4.5	1020.0	π-interaction with Phe338, dimer core
7	F328S	M-T	39.9 ± 0.4	81.0 ± 4.9	9.5	6740.0	π-interaction with Phe338, dimer core
8	F328 V	M-T	39.7 ± 0.2	115.8 ± 4.8	12.8	5860.0	π-interaction with Phe338, dimer core
9	T329I	M-T	73.2 ± 0.2	126.4 ± 2.7	-1.7	48.0	solvent-exposed
10	T329S	M-T	60.5 ± 0.2	108.7 ± 2.3	2.7	347.2	solvent-exposed
11	L330H	M-T	27.2 ± 0.5	103.4 ± 7.4	18.8	72100.0	center of the hydrophobic core, dimer core
12	L330P	M					center of the hydrophobic core, dimer core
13	L330R	M					center of the hydrophobic core, dimer core
14	Q331H	M-T	68.6 ± 0.2	149.6 ± 6.8	-0.1	22.6	intermonomer with 327
15	Q331P	M-T	60.2 ± 0.2	133.9 ± 4.8	3.4	137.8	intermonomer with 327
16	Q331R	M-T	72.7 ± 0.3	156.9 ± 6.6	-1.9	9.0	intermonomer with 327
17	I332 V	M-T	67.9 ± 0.2	151.3 ± 4.3	0.2	22.6	buried in the hydrophobic core, dimer core
18	G334 V	M-T-A	49.9 ± 0.2	128.5 ± 4.1	8.2	787.7	a tight turn
19	G334 W	M-T	53.0 ± 0.2	110.2 ± 2.5	5.8	777.2	a tight turn
20	R335G	M-T	46.4 ± 0.7	100.5 ± 26.5	8.2	2290.0	solvent-exposed
21	R335H	M-T	57.8 ± 0.2	118.5 ± 3.5	4.1	327.2	solvent-exposed
22	R335L	M-T	64.1 ± 0.2	137.9 ± 3.0	1.8	70.0	solvent-exposed
23	R337C	M-T-A	21.6 ± 0.7	92.0 ± 21.7	20.6	196000.0	salt-bridge with Asp352, hydrophobic core
24	R337L	M-T	36.9 ± 0.2	104.0 ± 3.7	13.2	10200.0	salt-bridge with Asp352, hydrophobic core
25	R337P	M	37.6 ± 0.4	81.5 ± 4.6	10.6	9200.0	salt-bridge with Asp352, hydrophobic core

Table 2.3 (continued)

No.	mutant	State	T_m (°C)	ΔH_u^{Tm} (kcal/mol)	$\Delta\Delta G_u^{Tm}$ (kcal/mol)	K_d (nM)	WT AA
26	F338I	M-T	36.8 ± 0.5	93.0 ± 7.0	12.1	10400.0	π-interaction with Phe328, dimer core
27	F338L	M-T	51.3 ± 0.1	103.3 ± 3.6	6.2	1140.0	π-interaction with Phe328, dimer core
28	E339 K	M-T	67.4 ± 0.2	134.9 ± 2.7	0.4	53.9	solvent-exposed
29	E339Q	M-T	66.4 ± 0.2	141.1 ± 4.9	0.8	45.2	solvent-exposed
30	F341C	M-D	23.8 ± 0.3	66.4 ± 3.3	15.4	64400.0	hydrophobic core, tetramer interface
31	R342L	M-T	62.4 ± 0.2	137.3 ± 4.9	2.5	90.2	dimer core, solvent-exposed
32	R342P	M					dimer core, solvent-exposed
33	R342Q	M-T	62.1 ± 0.2	134.9 ± 4.1	2.6	102.4	dimer core, solvent-exposed
34	E343G	M-T	57.9 ± 0.2	120.7 ± 4.0	4.1	302.3	mc interdimer H-bond with sc351
35	L344P	M					hydrophobic core, tetramer interface
36	L344R	M-D	39.0 ± 0.2	71.4 ± 2.8	9.0	7900.0	hydrophobic core, tetramer interface
37	E346A	M-T	64.6 ± 0.1	145.1 ± 3.4	1.7	47.7	solvent-exposed
38	A347G	M-T	55.3 ± 0.2	100.1 ± 2.8	4.4	793.4	tetramer interface
39	A347T	M-D	44.3 ± 0.4	62.2 ± 4.7	6.2	4967.5	tetramer interface
40	L348F	M-T	55.0 ± 0.2	128.6 ± 4.8	5.7	347.3	hydrophobic core, tetramer interface
41	L348S	M-T	32.6 ± 0.3	76.4 ± 4.6	12.3	18500.0	hydrophobic core, tetramer interface
42	E349D	M-T	54.3 ± 0.2	82.6 ± 2.6	4.1	1450.0	sc intermonomer H-bond with mc333
43	K351 N	M-T	54.0 ± 0.4	117.9 ± 3.1	5.7	549.5	solvent-exposed
44	D352H	M-T	60.6 ± 0.2	133.0 ± 4.3	3.3	138.1	forms a salt-bridge with Arg337
45	A353T	M-T	63.0 ± 03	128.5 ± 4.8	2.1	119.3	solvent-exposed
46	Q354E	M-T	59.3 ± 0.3	99.8 ± 4.1	2.9	539.6	solvent-exposed
47	Q354 K	M-T	64.1 ± 0.2	113.9 ± 3.1	1.5	197.3	solvent-exposed
48	Q354R	M-T	66.7 ± 0.2	117.2 ± 10.0	0.6	134.9	solvent-exposed
49	G356A	M-T	70.3 ± 0.2	152.4 ± 4.7	-0.9	15.6	solvent-exposed
50	G356 W	M-T	68.5 ± 0.2	145.0 ± 4.7	-0.1	28.6	solvent-exposed

^{a} The fraction of tetramer determined by gel filtration chromatography; asterisks indicate the fraction of dimer. M-T (Monomer–Tetramer); M-D (Monomer–Dimer); M (Monomer); T_m (transition temperature); ΔH_u^{Tm} (variation in the enthalpy of unfolding at T_m); $\Delta\Delta G_u^{Tm}$ (the difference in ΔG between WT and mutant peptides at the T_m of the WT peptide); sc (side chain); mc (main chain). The standard errors of fittings are indicated. The dissociation constant at 37 °C is calculated by $K_d = ((1 - K_u)/2)^{-1/3}$. For dimer mutants, we used $K_d = K_u^{-1}$

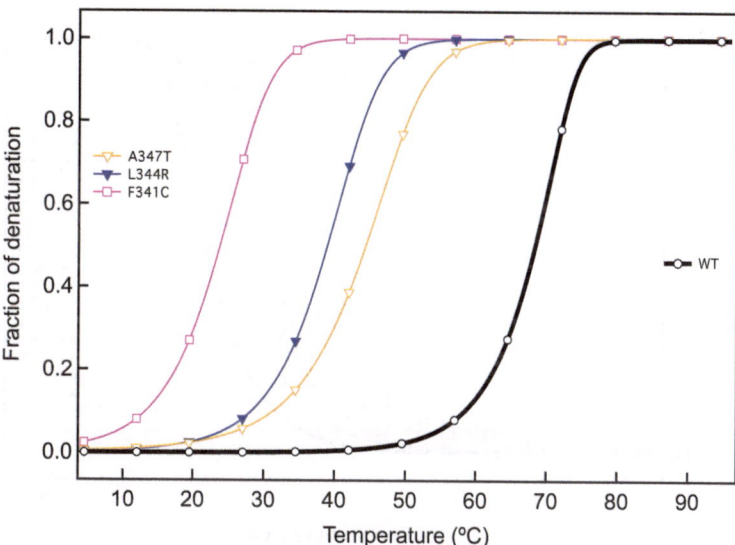

Fig. 2.6 Thermal denaturation of WT- and three dimer mutant-p53TD peptides (F341C, L344R, and A347T). Thermal denaturation of the peptides was analyzed by measuring the ellipticity at 222 nm for peptide solutions containing 10 μM peptide in 50 mM sodium phosphate, pH 7.5, 100 mM NaCl over the range of 4–96 °C, with a scan rate of 1 °C/min

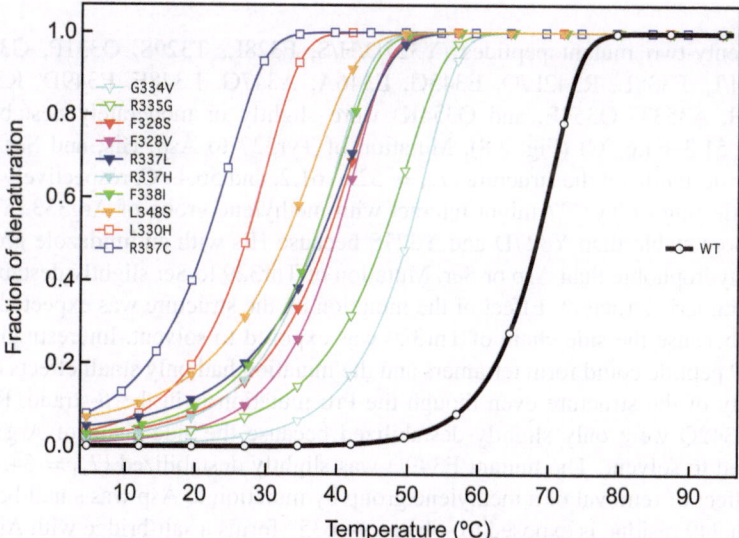

Fig. 2.7 Thermal denaturation of WT- and ten mutant-p53TD peptides (F328V, F328S, L330H, G334V, R335G, R337C, R337L, R337H, F338I, and L348S). Thermal denaturation of the peptides was analyzed by measuring the ellipticity at 222 nm for peptide solutions containing 10 μM peptide in 50 mM sodium phosphate, pH 7.5, 100 mM NaCl over the range of 4–96 °C, with a scan rate of 1 °C/min

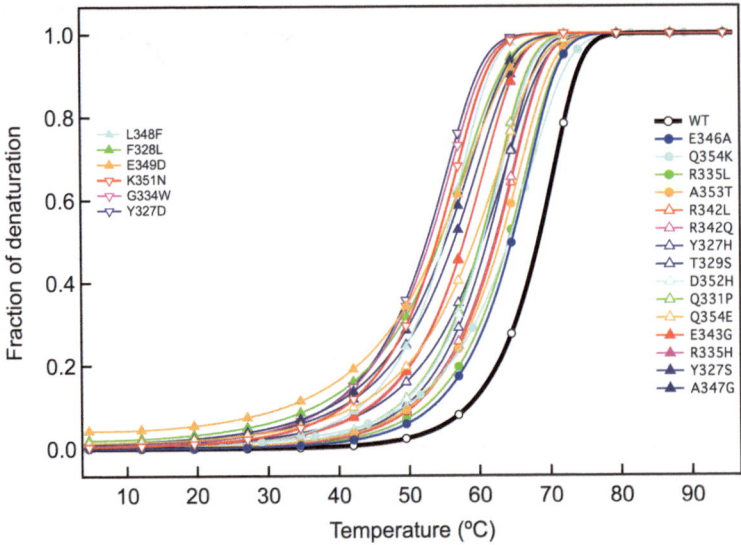

Fig. 2.8 Thermal denaturation of WT- and twenty-one mutant peptides (Y327D/H/S, F328L, T329S, Q331P, G334W, R335H/L, F338L, R342L/Q, E343G, E346A, A347G, L348F, E349D, K351 N, D352H, A353T, Q354E, and Q354K). Thermal denaturation of the peptides was analyzed by measuring the ellipticity at 222 nm for peptide solutions containing 10 μM peptide in 50 mM sodium phosphate, pH 7.5, 100 mM NaCl over the range of 4–96 °C, with a scan rate of 1 °C/min

Twenty-two mutant peptides (Y327D/H/S, F328L, T329S, Q331P, G334W, R335H/L, F338L, R342L/Q, E343G, E346A, A347G, L348F, E349D, K351N, D352H, A353T, Q354E, and Q354K) were slightly or moderately destabilized (T_m = 51.3–64.6 °C) (Fig. 2.8). Mutation of Tyr327 to Asp, His, and Ser moderately destabilized the structure (T_m = 52.6, 61.2, and 56.4 °C, respectively). The aromatic ring of Tyr327 might interact with methylene group of Arg333. Y327H was more stable than Y327D and Y327S because His with an imidazole group is more hydrophobic than Asp or Ser. Mutation of Thr329 to Ser slightly destabilized the tetrameric structure. Effect of the mutation on the structure was expected to be small because the side chain of Thr329 was exposed to solvent. Interestingly, the Q331P peptide could form tetramers and the mutation had only small effects on the stability of the structure even though the Pro mutation is in the β-strand. R342L and R342Q were only slightly destabilized because the side chain of Arg342 is exposed to solvent. The mutant E349D was slightly destabilized (T_m = 54.3 °C). The effect of removal of a methylene group by mutation to Asp was small because the Glu349 residue is exposed to solvent. Asp352 forms a salt-bridge with Arg337. Mutation of Asp352 to His moderately destabilized the structure, although mutation of Arg337 was strongly destabilizing. These results suggested that an electrostatic interaction such as a salt-bridge was less crucial to the structure than a hydrophobic interaction. A353T, Q354E, and Q354K slightly destabilized the tetrameric structure, because Ala353 and Gln354 are exposed to solvent.

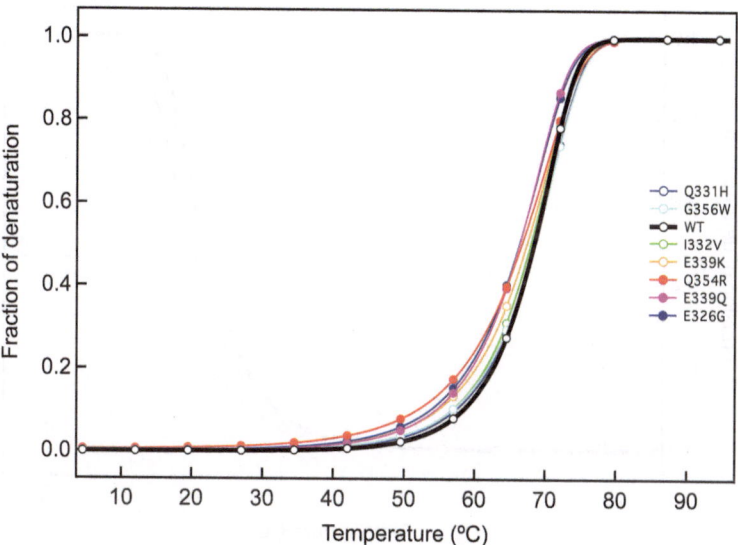

Fig. 2.9 Thermal denaturation of WT- and seven mutants (E326G, I332V, Q331H, E339K/Q, Q354R, and G356W). Thermal denaturation of the peptides was analyzed by measuring the ellipticity at 222 nm for peptide solutions containing 10 μM peptide in 50 mM sodium phosphate, pH 7.5, 100 mM NaCl over the range of 4–96 °C, with a scan rate of 1 °C/min

Seven mutants (E326G, I332V, Q331H, E339K/Q, Q354R, and G356W) showed almost the same stability as the WT (Fig. 2.9). Glu326 is located at the amino edge of a β-strand and exposed to the solvent. Mutation of Glu326 to Gly had little effect on the structure, although Gly has no side chain and does not tend to form a β-strand. I332V induced a limited change in the thermal stability of the tetrameric structures ($\Delta T_{\mathrm{m}} = -0.5$ °C, $\Delta\Delta G_{\mathrm{u}}^{Tm} = -0.2$ kcal/mol), even though Ile332 is buried at the hydrophobic core of the tetramer. Interestingly, the destabilization observed ($\Delta G_{\mathrm{u}}^{Tm} = -0.2$ kcal/mol) was smaller than the average found for the removal of a buried methylene group in monomeric proteins. E339K, E339Q, and G356W showed almost same stability as WT. Glu339 and Gly356 are exposed to solvent and thus, their mutations did not affect the stability of the tetramer.

Three mutants (T329I, Q331R, and G356A) were more stable than WT ($T_{\mathrm{m}} = 73.2$, 72.7, and 70.3 °C, respectively) (Fig. 2.10). Mutation of Glu331 to His or Arg slightly stabilized the tetrameric structure. It was expected that mutation of Gln331 would have little effect on the structure because the side chain of Gln331 is exposed to solvent. Intriguingly, Arg substitution for Gln331 caused an increase in stability of ≈4.3 °C. There are two possible reasons why Q331R stabilized the structure. One, the methylene group of Arg can stabilize the structure through hydrophobic interactions with amino acid residues around Gln331, such as the aromatic ring of Tyr327. Two, the cationic group of Arg can interact with the aromatic ring of Tyr327 by cation-π interaction. Gly356 is located at the carboxyl edge of an α-helix. Mutation of Gly356 to Ala results in a tendency to form an α-helix, which could in turn stabilize the tetrameric structure.

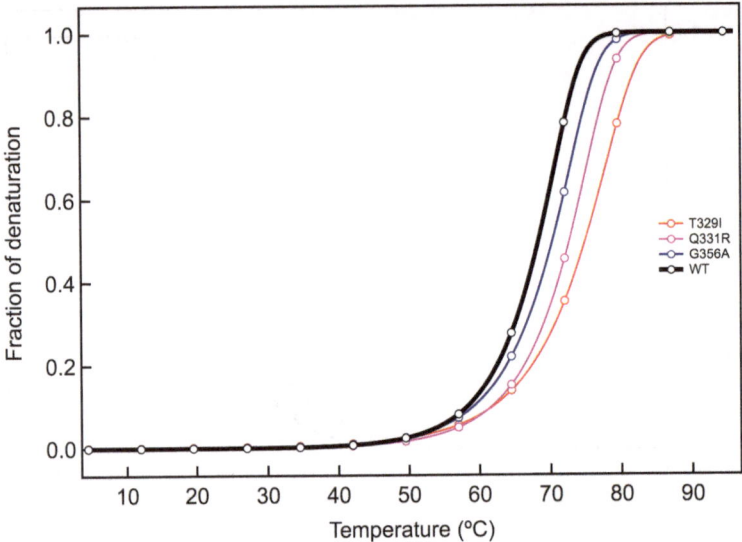

Fig. 2.10 Thermal denaturation of WT- and three mutants (T329I, Q331R, and G356A). Thermal denaturation of the peptides was analyzed by measuring the ellipticity at 222 nm for peptide solutions containing 10 μM peptide in 50 mM sodium phosphate, pH 7.5, 100 mM NaCl over the range of 4–96 °C, with a scan rate of 1 °C/min

2.3.5 Modeling of Mutant p53TD Peptides

Mutations that changed some solvent-exposed p53TD amino acid residues had little or no significant affect on the tetramers' thermal stability. To elucidate why these mutations occur in human cancers, the TD of each mutant were modeled (Fig. 2.12), finding that changes in some solvent-exposed residues altered the calculated electrostatic potential on the surface of the p53TD. This was especially so for E339K, E339Q, E343G, E346A, and Q354K. I suggest that these changes might influence either the interdomain or the intermolecular interactions with binding partners that thereby could account for their selection as cancer mutants.

2.3.6 Correlation Between Stability of p53TD Peptides and that of the Full-Length p53 Protein and the Transcriptional Activity

The stabilities of the tetrameric structures of the mutant p53TD peptides were compared with the oligomeric state of full-length p53-EGFP fusion proteins carrying TD mutations; FIDA was employed that yields a quantitative assessment of the fraction of protein oligomers in vivo at physiologically relevant concentrations

Fig. 2.11 Comparison of thermodynamic stability by CD and oligomeric status by gel filtration analysis. The T_m values of mutant peptides and the fraction of oligomers were obtained as shown in Table 2.2 and Table 2.3. The data for mutants are plotted as solid circles. Numbers refer to mutants as listed in Table 2.2. Monomer mutants 11, 12, 25, 32, and 35, and dimer mutants 30, 36, and 39 are not shown

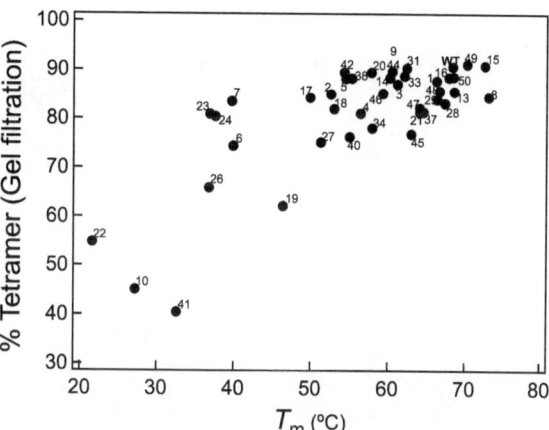

[28]. The clear correlation ($r = 0.77$) between the T_m measured here and the in vivo oligomerization state (Fig. 2.13) strongly suggests that the quantitative data on the tetrameric structure of p53 peptides is extendable to the full p53 protein.

Also, the stabilities of the tetrameric structures were compared with the transcriptional activity of full-length EGFP-p53 protein with TD mutations, which were analyzed by p53 reporter system in living cell [28]. There were strong correlation between the T_m and the in vivo transcriptional activity (Fig. 2.14).

2.4 Discussion

In this study, I represent the first comprehensive, quantitative biophysical analysis of the oligomeric state and thermal stability of the 50 TD mutants identified in human cancers. Most mutant p53TD peptides formed a WT-like tetrameric structure with diminished stability (Fig. 2.4). However, tetrameric mutants with altered hydrophobic core residues (F328, L330, R337, F338, F341, L344, and L348), except I322V, exhibited dramatic reductions in stability and, in some cases, unfolding of the peptide (e.g. L330H, P, R; R337C, P; R342P, L344P) as determined by CD measurements (Fig. 2.3). In particular, mutations that introduced proline in the α-helix devastated tetramer formation; some mutants could not form tetramers and existed as unfolded monomers (L330P/R, R337P, R342P, and L344P), or as folded dimers (F341C, L344R, and A347T). Indeed, the thermal denaturation study predicted that several TD mutants (e.g. L330H, R337C/H/L, F338I, F341C, L344R, and L348S) are thermally unstable at or near body temperature. These results are consistent with an alanine-scanning study of the p53TD that identified nine key hydrophobic resides important for TD thermal stability and oligomerization [18].

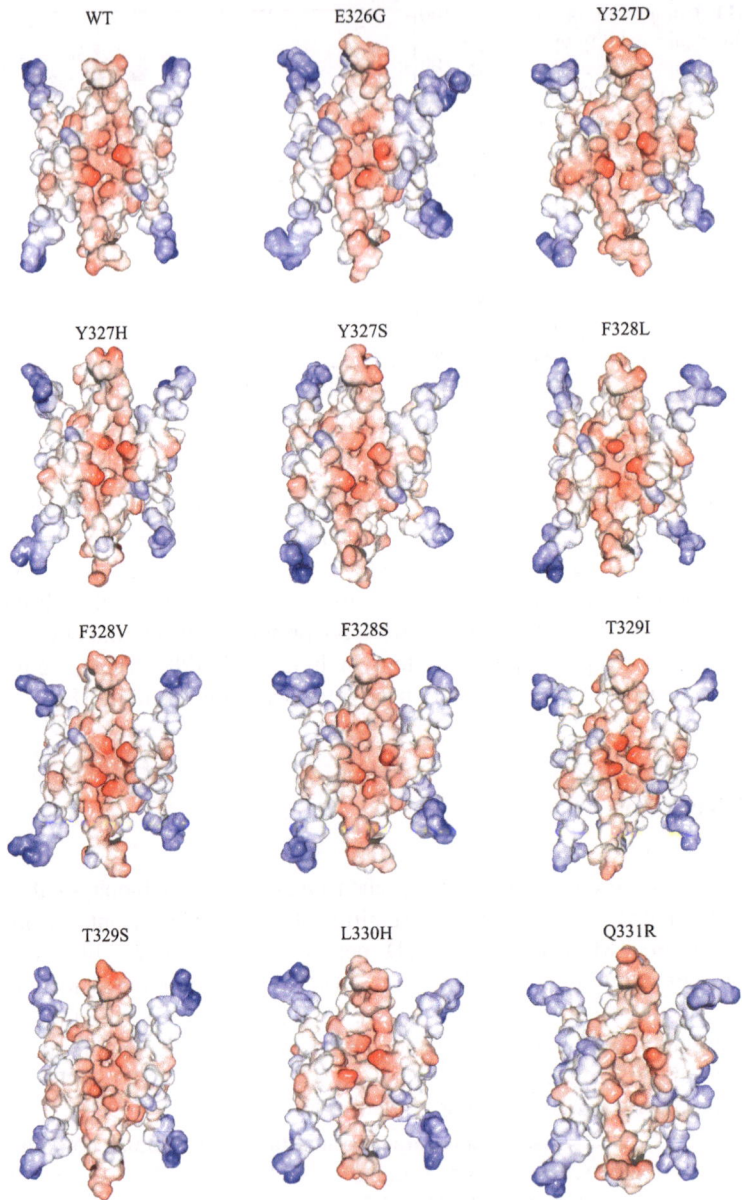

Fig. 2.12 The surface modeling of *mutant p53TDs*. A solvent exposed surface of the p53TD mutants, colored by electrostatic potential. The positively charged surface in *blue* and negative in *red*

In contrast to mutations that affect hydrophobic core residues, mutations that affect residues accessible to solvent were less destabilizing. The WT p53TD is thermally quite stable ($T_m \sim 70$ °C, Table 2.3) compared with the core DNA-binding domain

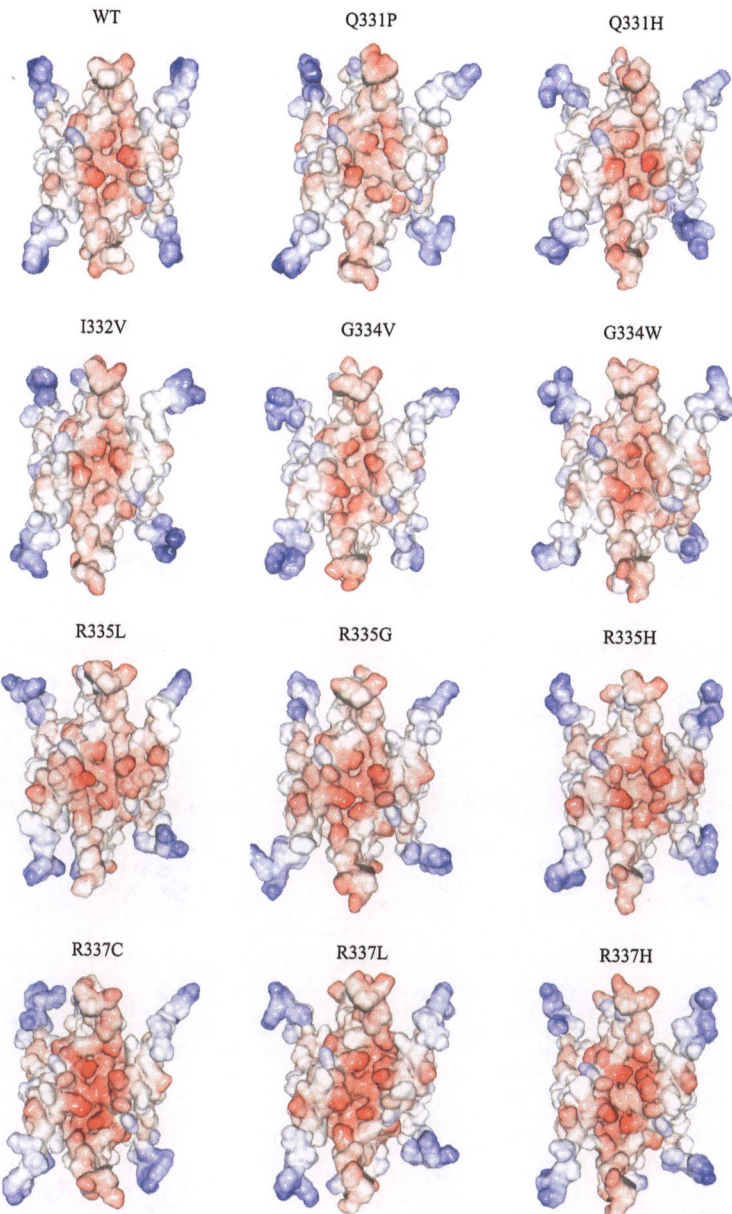

Fig. 2.12 continued

[15]; most mutations that affect TD domain residues, except as noted above, would not be expected to unfold the domain structure. Nevertheless, for most thermally stable TDs, the change in amino acids significantly altered the disassociation constant for tetramer formation (Table 2.3). Importantly, because tetramers are essential for DNA

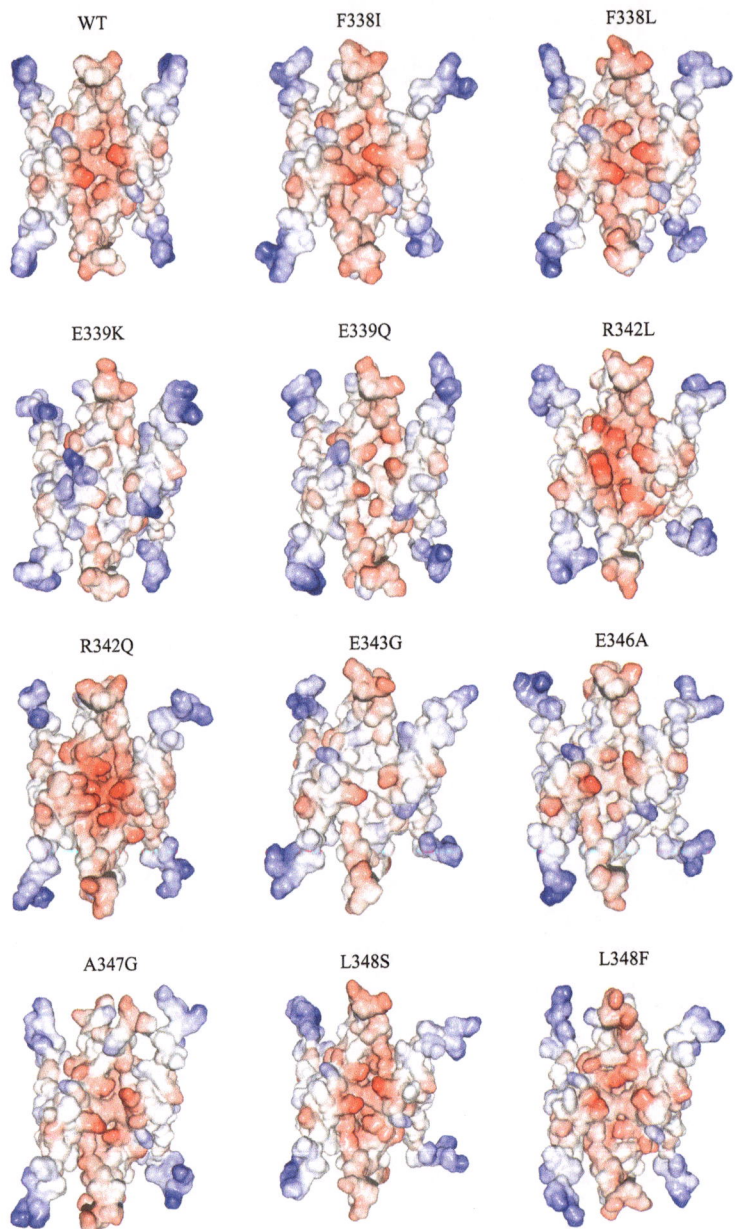

Fig. 2.12   continued

binding and activating transcription [9], and the p53 tetramer is in equilibrium with the monomer, the intra-nuclear p53 concentration is a critical factor in determining p53 function. In cultured, undamaged human embryonic skin fibroblasts (WS-1 cells), p53

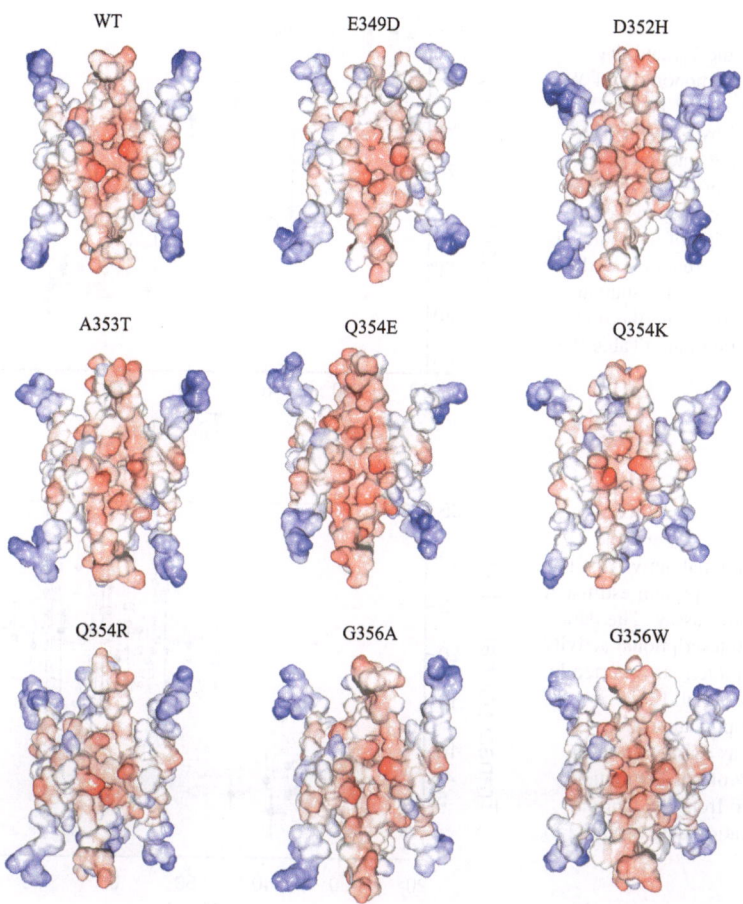

WT   E349D   D352H

A353T   Q354E   Q354K

Q354R   G356A   G356W

Fig. 2.12 continued

abundance was 7.8×10^3 molecules/cell, whereas after DNA damage from neo-carzinostatin, it rose $\sim$ threefold to 21.8×10^3 molecules/cell [35]. The volume of the nucleus of a human fibroblast is about 10^{-12} L [36]. Correspondingly, the p53 concentration in the nucleus of normal, unstressed human cells is ≈ 13 nM and increases to ≈ 36 nM after DNA damage. At these concentrations, the fraction of tetramer for TD mutants at 37 °C can be assessed (Fig. 2.15). Thus, I predict that about 80 % of the accumulated WT p53 protein (WT p53TD $K_d = 10.2$ nM) is in the tetrameric state following DNA damage. These values might be somewhat high as the skin fibroblasts were cultured under normoxic conditions (~ 21 % oxygen), i.e., much higher that the oxygen concentration in most tissues (~ 5–8 %). Oxidative stress activates p53, and normoxia causes oxidative stress in cultured cells [32]. Nevertheless, the dissociation constant for the formation of the WT p53 tetramer seems tuned to accommodate a much greater change in p53 function than the $\sim$ threefold change in nuclear

Fig. 2.13 Correlation between the T_m and the fraction of monomers of WT and mutant p53 protein estimated by FIDA. FIDA data for the fraction of monomer of p53 protein as reported by Imagawa et al. [28]. Fraction of monomers is plotted as a function of the stability of p53TD mutants obtained from the thermal denaturation data (Table 2.3)

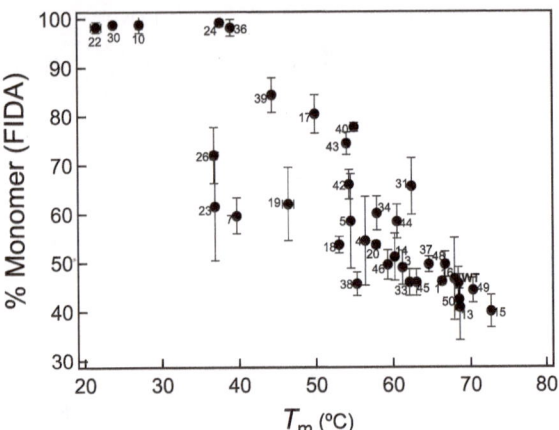

Fig. 2.14 Correlation between the T_m and the transcriptional activity of the mutant p53 protein estimated by reporter assay. The data for the transcriptional activity of p53 protein as reported by Imagawa et al. [28]. Transcriptional activity is plotted as a function of the stability of p53TD mutants obtained from the thermal denaturation data (Table 2.3)

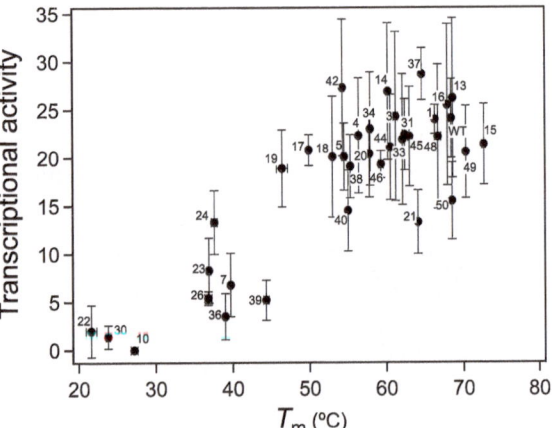

concentration. Hence, cancer mutations that affect the p53TD K_d by more than $\sim$ 10-fold might not elicit a sufficient concentration of tetramers in cells to induce the transcriptional responses important for tumor suppression [3]. Additionally, p53-mediated transcription-independent apoptosis, which reportedly depends on the ability of p53 to form tetramers in the cytoplasm [12], might be diminished. Nevertheless, I argue that even mild changes to the K_d for tetramer formation or in p53 stability could significantly affect p53 function because of the sensitivity of tetramer formation to the K_d and p53 concentration in the physiological range. Many previous studies involved a $\sim$ 20-fold p53 overexpression (Fig. 2.16) that would not reveal the detrimental effects of many p53TD mutations (Fig. 2.17). These results suggested that care should be taken when p53 function is analyzed under nonnative conditions, such as in a transient expression system, where the protein concentration is very high.

Fig. 2.15 Fraction of
tetramer at 37 °C at
concentrations of 13 nM
(endogenous p53 level in
unstressed cell) and 36 nM
(stressed cell) against the
value of K_d. Each data point
represents the value of a
mutant at 13 nM (solid
circles) and 36 nM (open
circles). Monomer mutants
11, 12, 25, 32, and 35, and
dimer mutants 30, 36, and 39
are not shown

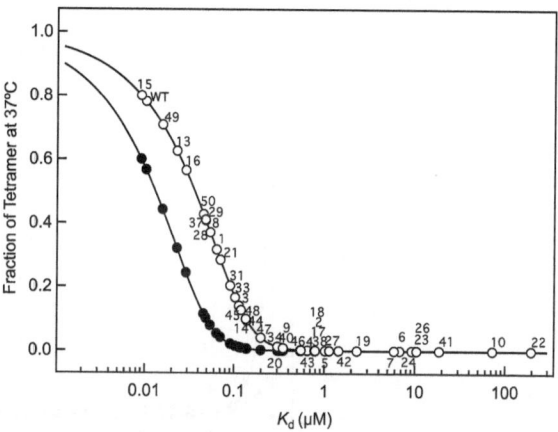

In addition to the direct effects of mutations on tetramer formation, several indirect effects may further acerbate p53 function (Fig. 2.18). First, the p53TD contains a nuclear export signal (NES, M340-K351) that is exposed in the monomer and dimer but not in the tetramer [33]. Except for those mutants in the α-helix between M340 and K351 that affect the interaction of the NES with CRM1 [33], TD mutants potentially could acerbate tetramer formation by increasing the cytoplasmic export of p53 and preventing its nuclear accumulation to normal levels. Second, several post-translational modifications that potentially modulate p53 activity are influenced by its oligomeric state, and the oligomeric state can be modulated by post-translational modifications. The phosphorylation of Ser392 enhances tetramer formation [30]. The p53TD peptide in which His replaced L330 was thermally destabilized (Table 2.3), and the phosphorylation of p53 Ser392 by casein kinase 2 was diminished when L330 was mutated to His [37]. Deletion of the residues 334–354 abolishes the ability of Chk1 to phosphorylate p53 [8]; sites that Chk1 can phosphorylate are believed to modulate p53 activity and stability [4]. The PCAF acetyltransferase, which acetylates Lys320, specifically recognizes p53's tetrameric structure [6]. In addition, Pirh2, an E3 ubiquitin protein-ligase that binds and ubiquitylates p53 protein in vitro and in vivo, only acts on its tetrameric form [38].

More than 50 proteins reportedly interact with the C-terminal region of the p53 protein, and several of them either require or influence tetramer formation. Tetramer formation of p53 is essential for its interaction with HPV-16 E2, c-Abl, and Mdm2 [9, 21]. The binding affinity of p53 to MDM2 fell when p53 contained the mutation L344P or R337C found in Li-Fraumeni patients [21]. c-Abl binds directly to the C-terminal basic domain of p53, and this interaction requires a tetramer. c-Abl may stabilize the tetrameric conformation, resulting in a more stable p53-DNA complex [39]. In contrast, the interaction of ARC with the p53TD inhibits tetramer formation and increases nuclear export [40]. The binding of S100 family proteins depends on the oligomeric status of p53 and controls the balance between monomer and tetramer [41]. Binding of the 14-3-3 protein to p53 enhances sequence-specific DNA binding by inducing p53 to form tetramers at lower concentrations [42].

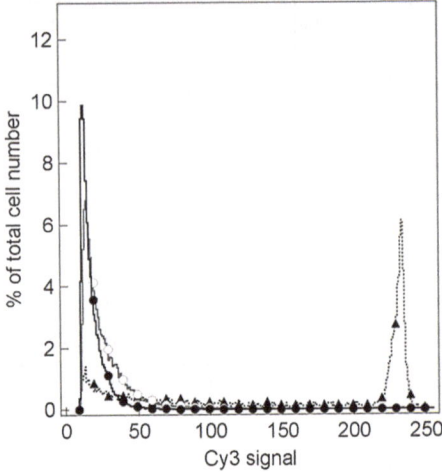

Fig. 2.16 Fraction of tetramer at 37 °C in cells when overexpressed. H1299 cells (p53 null) were transiently transfected p53 expression vector. The H1299 cells and A549 cells (expressed wild-type p53) with or without UV (25 J/m^2) treatment were stained with anti-p53 monoclonal antibody (DO-1) and Cy3-conjugated secondary antibody. The Cy3 signal intensity of each cell was measured and the distributions of Cy3 signal intensity are shown; A549 (endogenous p53 level without UV; *solid circle, solid line*), A549 (endogenous p53 level with UV; *open circle, solid gray line*), and H1299 (transient transfection; *solid triangle, dotted line*). The detailed methods were described as previously [28]

Fig. 2.17 Fraction of tetramer at 37 °C in cells when overexpressed. **b** Each data point represents the value of a mutant at 36 nM, endogenous p53 level in stressed cell, (open circles) and 720 nM, p53 level in cell when overexpressed, (*solid triangles*). Monomer mutants 11, 12, 25, 32, and 35, and dimer mutants 30, 36, and 39 are not shown

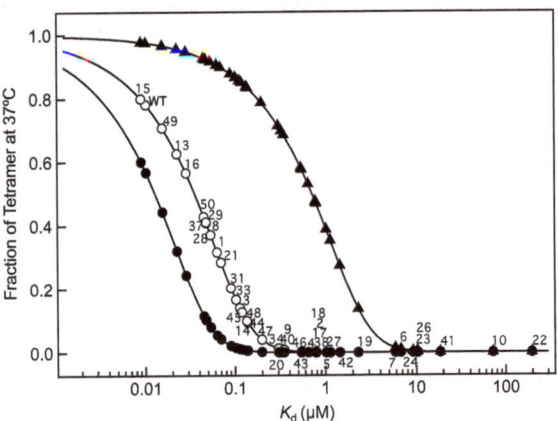

The p53TD from ∼ 13 apparently cancer-associated mutants in eight residues, mostly in the α-helical region of the TD, only moderately affected, by ∼fivefold, the K_d of tetramer formation of E326G, T329I, R335L, E339 K, E339Q, and E346A, and very slightly affected, by twofold, that of Q331H, Q331R, I332 V, G356A and G356 W (Table 2.3). The apparently minimal effect of these changes

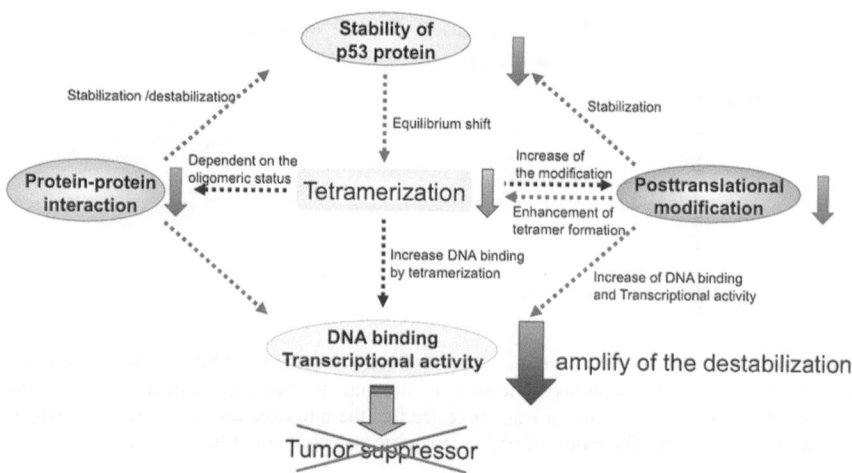

Fig. 2.18 Amplification of the destabilization effects of mutations. Tumor suppressor activity of p53 is regulated by many factors, including tetramerization, posttranslational modification, protein–protein interactions. Mutations of tetramerization domain directly destabilize the tetrameric structure and decrease p53 function (*dark gray arrow*). Because tetramer formation is crucial for posttranslational modification, DNA binding, and protein–protein interactions, destabilization of the tetrameric structure decreases them. Thus, mutations of tetramerization domain indirectly affect the posttranslational modification and protein–protein interactions and result in more destabilization of the structure and decrease function

is particularly surprising for mutations that affect E326, I332, E339, and E346, because these are among the 12 most highly evolutionarily conserved residues in the TD [9], and changes to conserved residues often are deleterious to function. I questioned why then do the mutations causing these changes exist among p53-associated cancer mutants? As Soussi et al. noted [43], mistakes occur in the literature on p53, possibly due to errors in sequencing or PCR, so caution is needed about accepting mutants that have been reported in cancers only once or a few times; data on germline mutants should be more reliable. Of the 13 mutants noted above, all but four (Q331H, Q331R, R342Q, G356W) occur only once in the IARC TP53 mutant database (http://www-p53.iarc.fr/), and none have been reported as germline mutations. Thus, some of these mutants may be false reportings. Of the remaining four mutants (Fig. 2.12), two clearly are solvent exposed residues that either change the surface charge (R342Q) or replace a small residue with one bearing a bulky, hydrophobic side chain at the surface (G356W). Although mutations affecting solvent exposed residues and altering the electrostatic potential of p53′s surface were less thermally destabilizing than core mutants were, the change in surface charge potential might well affect intra-protein interactions or the interaction of the TD with one or more of its many binding partners. R342Q represents such a change, and the G356W change might disrupt surface complementarity that could affect protein interactions important for p53′s function. Many mutants with greatly changed K_ds also involve surface-exposed alterations in

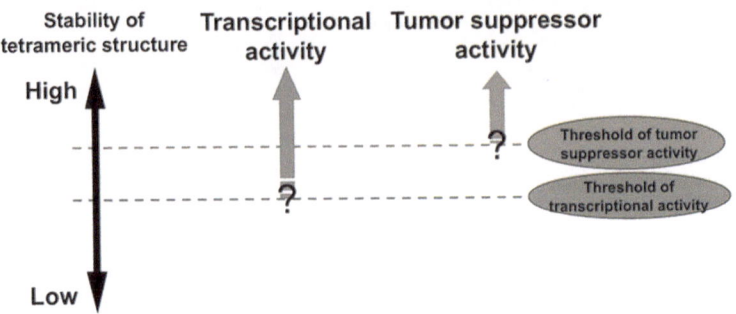

Fig. 2.19 Threshold for loss of tumor suppressor activity of p53 in terms of the disruption of tetrameric structure. The stability of tetrameric structure is correlated with the transcriptional activity of p53. From our results, it was suggested that the threshold for loss of tumor suppressor activity in terms of the disruption of p53′s tetrameric structure could be extremely low

charge that affect the predicted electrostatic potential of the p53TD′s surface (Fig. 2.3). In the crystal structure, E349D, a change that moderately increases the K_d (to 1450 nM) is implicated in crystal contacts and, therefore, probably is important for such interactions [44].

Residue Q331 is in the short β-sheet that forms part of the monomer–monomer interface, but Q331 is not involved in monomer–monomer interactions, and the change to either His, Arg, or even Pro had only minor effects on p53TD thermal stability (Table 2.3). A recent yeast-based assay for transcriptional activation revealed that almost any amino acid sufficed at this position [45]. Thus, biophysical or biochemical measurements do not show why mutations that alter this residue appear among cancer-associated ones.

Analyses of SNPs in p53 and its pathway support the suggestion of the potential importance of the subtle effects of TD mutants on tetramer formation. The *TP53* gene reportedly contains nineteen exonic polymorphisms, among which researchers have validated four (R47S, R72P, V217M, and G360A). The codon 47P/S and 72R/P polymorphisms subtly alter expression of p53 transcriptional targets. Although controversial [46], molecular evidence suggests that both polymorphisms alter cancer risk [47–49]. Additional evidence comes from SNPs in p53 pathways. Bond et al., working on the most intensively studied T/G SNP at nucleotide 309 in the first intron of the *MDM2* gene, demonstrated that the 309G variant is bound more efficiently by the transcription factor SP1, thereby increasing the efficiency of synthesizing MDM2, and consequently, slightly lowering levels of p53 [50]. The estrogen receptor also binds the *MDM2* promoter in the region of SNP309 and also can increase MDM2 expression in response to the hormone. Several studies associated MDM2 SNP309 polymorphism and increased cancer risk in males and females, although others saw no such connection [46]. Thus, although further work is required, these studies support the hypothesis that relatively small changes in p53 concentration or the ability to form tetramers could contribute to cancer risk or progression. I suggest that the threshold for loss of tumor suppressor activity in terms of the disruption of p53's tetrameric structure could be extremely low (Fig. 2.19). Furthermore, the study of cancer-associated mutants in the TD that minimally affect tetramer formation may reveal additional functions for this domain in p53 biology. The main result in this chapter was published first in *Journal of Biological Chemistry* in 2011, and this chapter is its expanded version [51].

References

1. Hanahan D, Weinberg RA (2000) The hallmarks of cancer. Cell 100:57–70
2. Halazonetis TD, Gorgoulis VG, Bartek J (2008) An oncogene-induced DNA damage model for cancer development. Science 319:1352–1355
3. Zilfou JT, Lowe SW (2009) Tumor suppressive functions of p53. Cold Spring Harb Perspect Biol 2:a000935–a000935
4. Meek DW, Anderson CW (2009) Posttranslational modification of p53: cooperative integrators of function. Cold Spring Harb 1–16: 528–536
5. Halazonetis TD, Kandil AN (1993) Conformational shifts propagate from the oligomerization domain of p53 to its tetrameric DNA binding domain and restore DNA binding to select p53 mutants. EMBO J 12:5057–5064
6. Sakaguchi K, Herrera JE, Saito S, Miki T, Bustin M, Vassilev A, Anderson CW, Appella E (1998) DNA damage activates p53 through a phosphorylation-acetylation cascade. Genes Dev 12:2831–2841
7. Maki CG (1999) Oligomerization is required for p53 to be efficiently ubiquitinated by MDM2. J Biol Chem 274:16531–16535
8. Shieh SY, Ahn J, Tamai K, Taya Y, Prives C (2000) The human homologs of checkpoint kinases Chk1 and Cds1 (Chk2) phosphorylate p53 at multiple DNA damage-inducible sites. Genes Dev 14:289–300
9. Chene P (2001) The role of tetramerization in p53 function. Oncogene 20:2611–2617
10. Warnock LJ, Knox A, Mee TR, Raines SA, Milner J (2008) Influence of tetramerisation on site-specific post-translational modifications of p53: comparison of human and murine p53 tumor suppressor protein. Cancer Biol Ther 7:1481–1489
11. Itahana Y, Ke H, Zhang Y (2009) p53 Oligomerization is essential for its C-terminal lysine acetylation. J Biol Chem 284:5158–5164
12. Pietsch EC, Perchiniak E, Canutescu AA, Wang G, Dunbrack RL, Murphy ME (2008) Oligomerization of BAK by p53 utilizes conserved residues of the p53 DNA binding domain. J Biol Chem 283:21294–21304
13. Hainaut P, Hollstein M (2000) p53 and human cancer: the first ten thousand mutations. Adv Cancer Res 77:81–137
14. Petitjean A, Mathe E, Kato S, Ishioka C, Tavtigian SV, Hainaut P, Olivier M (2007) Impact of mutant p53 functional properties on TP53 mutation patterns and tumor phenotype: lessons from recent developments in the IARC TP53 database. Hum Mutat 28:622–629
15. Joerger AC, Fersht AR (2010) The tumor suppressor p53: from structures to drug discovery. Cold Spring Harb Perspect 2(6):a000919–a000919
16. Clore GM, Ernst J, Clubb R, Omichinski JG, Kennedy WMP, Sakaguchi K, Appella E, Gronenborn AM (1995) Refined solution structure of the oligomerization domain of the tumour suppressor p53. Nat Struct Biol 2:321–333
17. Jeffrey PD, Gorina S, Pavletich NP (1995) Crystal structure of the tetramerization domain of the p53 tumor suppressor at 1.7 angstroms. Science 267:1498–1502

18. Mateu MG, Fersht AR (1998) Nine hydrophobic side chains are key determinants of the thermodynamic stability and oligomerization status of tumour suppressor p53 tetramerization domain. EMBO J 17:2748–2758
19. DiGiammarino EL, Lee AS, Cadwell C, Zhang W, Bothner B, Ribeiro RC, Zambetti G, Kriwacki RW (2002) A novel mechanism of tumorigenesis involving pH-dependent destabilization of a mutant p53 tetramer. Nat Struct Biol 9:12–16
20. Davison TS, Yin P, Nie E, Kay C, Arrowsmith CH (1998) Characterization of the oligomerization defects of two p53 mutants found in families with Li-Fraumeni and Li-Fraumeni-like syndrome. Oncogene 17:651–656
21. Lomax ME, Barnes DM, Hupp TR, Picksley SM, Camplejohn RS (1998) Characterization of p53 oligomerization domain mutations isolated from Li-Fraumeni and Li-Fraumeni like family members. Oncogene 17:643–649
22. Atz J, Wagner P, Roemer K (2000) Function, oligomerization, and conformation of tumor-associated p53 proteins with mutated C-terminus. J Cell Biochem 76:572–584
23. Rollenhagen C, Chene P (1998) Characterization of p53 mutants identified in human tumors with a missense mutation in the tetramerization domain. Int J Cancer 78:372–376
24. Johnson CR, Morin PE, Arrowsmith CH, Freire E (1995) Thermodynamic analysis of the structural stability of the tetrameric oligomerization domain of p53 tumor suppressor. Biochemistry 34:5309–5316
25. Maki CG, Huibregtse JM, Howley PM (1996) In vivo ubiquitination and proteasome-mediated degradation of p53(1), Cancer Res 56:2649–2654
26. Kato S, Han SY, Liu W, Otsuka K, Shibata H, Kanamaru R, Ishioka C (2003) Understanding the function-structure and function-mutation relationships of p53 tumor suppressor protein by high-resolution missense mutation analysis. Proc Natl Acad Sci USA 100:8424–8429
27. Kawaguchi T, Kato S, Otsuka K, Watanabe G, Kumabe T, Tominaga T, Yoshimoto T, Ishioka C (2005) The relationship among p53 oligomer formation, structure and transcriptional activity using a comprehensive missense mutation library. Oncogene 24:6976–6981
28. Imagawa T, Terai T, Yamada Y, Kamada R, Sakaguchi K (2009) Evaluation of transcriptional activity of p53 in individual living mammalian cells. Anal Biochem 387:249–256
29. Nomura T, Kamada R, Ito I, Chuman Y, Shimohigashi Y, Sakaguchi K (2009) Oxidation of methionine residue at hydrophobic core destabilizes p53 tetrameric structure. Biopolymers 91:78–84
30. Sakaguchi K, Sakamoto H, Lewis MS, Anderson CW, Erickson JW, Appella E, Xie D (1997) Phosphorylation of serine 392 stabilizes the tetramer formation of tumor suppressor protein p53. Biochemistry 36:10117–10124
31. Šli A, Blundell TL (1993) Comparative protein modelling by satisfaction of spatial restraints. J Mol Biol 234:779–815
32. Parrinello S, Samper E, Krtolica A, Goldstein J, Melov S, Campisi J (2003) Oxygen sensitivity severely limits the replicative lifespan of murine fibroblasts. Nat Cell Biol 5:741–747
33. Stommel JM, Marchenko ND, Jimenez GS, Moll UM, Hope TJ, Wahl GM (1999) A leucine-rich nuclear export signal in the p53 tetramerization domain: regulation of subcellular localization and p53 activity by NES masking. EMBO J 18:1660–1672
34. Fernandez-Fernandez MR, Veprintsev DB, Fersht AR (2005) Proteins of the S100 family regulate the oligomerization of p53 tumor suppressor. Proc Natl Acad Sci USA 102:4735–4740
35. Wang YV, Wade M, Wong E, Li YC, Rodewald LW, Wahl GM (2007) Quantitative analyses reveal the importance of regulated Hdmx degradation for p53 activation. Proc Natl Acad Sci USA 104:12365–12370
36. Swanson JA, Lee M, Knapp PE (1991) Cellular dimensions affecting the nucleocytoplasmic volume ratio. J Cell Biol 115:941–948

37. Chene P (2000) Fast, qualitative analysis of p53 phosphorylation by protein kinases. Biotechniques 28:240–242
38. Sheng Y, Laister RC, Lemak A, Wu B, Tai E, Duan S, Lukin J, Sunnerhagen M, Srisailam S, Karra M, Benchimol S, Arrowsmith CH (2008) Molecular basis of Pirh2-mediated p53 ubiquitylation. Nat Struct Mol Biol 15:1334–1342
39. Nie Y, Li HH, Bula CM, Liu X (2000) Stimulation of p53 DNA binding by c-Abl requires the p53 C terminus and tetramerization. Mol Cell Biol 20:741–748
40. Foo RSY, Nam YJ, Ostreicher MJ, Metzl MD, Whelan RS, Peng CF, Ashton AW, Fu W, Mani K, Chin SF, Provenzano E, Ellis I, Figg N, Pinder S, Bennett MR, Caldas C, Kitsis RN (2007) Regulation of p53 tetramerization and nuclear export by ARC. Proc Natl Acad Sci USA 104:20826–20831
41. van Dieck J, Fernandez–Fernandez MR, Veprintsev DB, Fersht AR (2009) Modulation of the oligomerization state of p53 by differential binding of proteins of the S100 family to p53 monomers and tetramers. J Biol Chem 284:13804–13811
42. Rajagopalan S, Jaulent AM, Wells M, Veprintsev DB, Fersht AR (2008) 14-3-3 activation of DNA binding of p53 by enhancing its association into tetramers. Nucleic Acids Res 36: 5983–5991
43. Soussi T, Kato S, Levy PP, Ishioka C (2005) Reassessment of the TP53 mutation database in human disease by data mining with a library of TP53 missense mutations. Hum Mutat 25:6–17
44. Miller M, Lubkowski J, Rao JKM, Danishefsky AT, Omichinski JG, Sakaguchi K, Sakamoto H, Appella E, Gronenborn AM, Clore GM (1996) The oligomerization domain of p53: crystal structure of the trigonal form. FEBS Lett 399:166–170
45. Merritt J, Roberts KG, Butz JA, Edwards JS (2007) Parallel analysis of tetramerization domain mutants of the human p53 protein using PCR colonies. Genomic Med 1:113–124
46. Whibley C, Pharoah PD, Hollstein M (2009) p53 polymorphisms: cancer implications. Nat Rev Cancer 9:95–107
47. Feng L, Hollstein M, Xu Y (2006) Ser46 phosphorylation regulates p53-dependent apoptosis and replicative senescence. Cell Cycle 5:2812–2819
48. Mantovani F, Tocco F, Girardini J, Smith P, Gasco M, Lu X, Crook T, Del Sal G (2007) The prolyl isomerase Pin1 orchestrates p53 acetylation and dissociation from the apoptosis inhibitor iASPP. Nat Struct Mol Biol 14:912–920
49. Bergamaschi D, Samuels Y, Sullivan A, Zvelebil M, Breyssens H, Bisso A, Del Sal G, Syed N, Smith P, Gasco M, Crook T, Lu X (2006) iASPP preferentially binds p53 proline-rich region and modulates apoptotic function of codon 72-polymorphic p53. Nat Genet 38:1133–1141
50. Bond GL, Hu W, Bond EE, Robins H, Lutzker SG, Arva NC, Bargonetti J, Bartel F, Taubert H, Wuerl P, Onel K, Yip L, Hwang SJ, Strong LC, Lozano G, Levine AJ (2004) A single nucleotide polymorphism in the MDM2 promoter attenuates the p53 tumor suppressor pathway and accelerates tumor formation in humans. Cell 119:591–602
51. Kamada R, Nomura T, Anderson CW, Sakaguchi K (2011) Cancer-associated p53 tetramerization domain mutants: quantitative analysis reveals a low threshold for tumor suppressor inactivation. J Biol Chem 286:252–258

Chapter 3
Stabilization of Mutant Tetrameric Structures by Calixarene Derivatives

3.1 Introduction

Li-Fraumeni syndrome (LFS) is a familial clustering of early onset tumors including sarcomas, breast cancer, brain tumors and adrenocortical carcinomas (ADC) [1]. The *TP53* and *CHEK2* genes are associated with Li-Fraumeni syndrome. More than half of all families with Li-Fraumeni syndrome have inherited mutations in the *TP53* gene. In contrast to other tumor suppressor genes that are mainly altered by truncating mutations, the majority of TP53 mutations are missense mutations (75 %). Missense mutations located within the DNA-binding loops of the protein were found associated with a higher prevalence of brain tumors and an earlier age at onset of breast cancers. In contrast, mutations outside the DNA binding loops were associated with a higher prevalence of ADC. R337H (CGC to CAC), located within the tetramerization domain of p53 has been first reported in Brazilian children with ADC [2]. R337H carriers have a high prevalence of ADC, breast cancers, and sarcomas. Indeed, R337H is the most frequent germline mutation for the *TP53* gene. Structural study has identified that R337H mutant had a pH-dependent instability of the mutant p53 tetramer in the physiological range [3]. This sensitivity correlates with the protonation state of the mutated His337 residue. In functional studies, p53 protein with the R337H mutation showed defect in oligomerization and decreased the transcriptional activity [4, 5].

The tumor suppressor protein p53 is a 393 amino acid phosphoprotein that is composed of five domains: an N-terminal transactivation domain, a proline rich domain, a central site-specific DNA binding domain, a tetramerization domain (TD), and a C-terminal basic domain. p53 controls apoptosis and cell cycle arrest in response to DNA damage through inducing or repressing the transcription of

R. Kamada, *Tetramer Stability and Functional Regulation of Tumor Suppressor Protein p53*, Springer Theses, DOI: 10.1007/978-4-431-54135-6_3, © Springer Japan 2012

several genes. Many structural and functional studies have revealed that the tet-ramer formation of p53 is essential for its tumor suppressor activity. The tetra-merization domain consists of a β-strand (Glu326-Arg333), a tight turn (Gly334), and an α-helix (Arg335-Gly356) [6]. Two monomers form a primary dimer via antiparallel β-sheet and α-helices, and the two primary dimers associate into a tetramer with an unusual four-helix-bundle [7, 8]. In the tetrameric structure, the guanidinium group of R337 side-chain forms a salt bridge (plus hydrogen-bond) with the carboxy group of D352 side-chain. Moreover, the methylene group from the R337 side-chain stabilizes the tetrameric structure via hydrophobic interactions.

Mutation of DNA binding domain can be mainly classified into two groups, a DNA contact mutation and a structural mutation. There are six mutation hot spots cluster in the DNA binding domain: two DNA contact mutation (R248 and R273) and four structural mutations (R175, G245, R249, and R282). These hot spots mutants can not bind a p53-specific promoter sequences (two repeats of PuPu-PuC(A/T)(A/T)GPyPyPy, in which Pu is a purine and Py is a pyrimidine). Structural mutants can be rescued by stabilization of their thermodynamic sta-bility. One of the structural mutants in the DNA binding domain, V143A, has been rescued by increasing the stability of the p53 core domain by 2 kcal/mol via two amino acid substitution (N239Y and N268D). Foster et al. [9] reported that small molecules, CP-257042 and CP-31398, rescued mutant p53 function by stabilizing a native conformation of DNA core domain. In the presence of the two compounds, core domain mutants (V173A, R175S, R249S, and R273H) bound PAb1620, which can only recognize native conformation of the DNA core domain. CP-31398 activates p53 transcription in V173A or R249S. Furthermore, CP31398 suppress the growth of colon carcinoma cells by 75 % in mice that harbor tumors. There is another small molecule which rescues mutant p53 function in living cancer cells, PRIMA-1, but its mechanism of action was unknown [10]. Fersht et al. [11] reported that a peptide, CDB3 (REDEDEIEW), could bind to and stabilize the p53 core domain in vitro. For DNA contact mutants, R273H, which retains an intact native fold, can be activated by C-terminal binding peptide via stabilization of the p53 tetramer [12]. For tetra-merization domain mutants, destabilization of the tetrameric structure causes dysfunction of p53 activity. Recently, Gordo et al. [13] reported that guanidi-nium-calix[4]arene which have four guanidinium groups could stabilize the tet-rameric structure of mutant R337H and G334V. HSQC-NMR analysis has revealed that the compound can stabilize the mutant tetrameric structure through interaction with a hydrophobic pocket and Glu residue (Glu336 and Glu339) on the tetrameric surface of p53. However, this compound could not stabilize the p53TD in physiological pH and salt concentration. In this study, we report sta-bilization of the mutant R337H tetrameric structure in physiological conditions and rescue of the transcriptional activity of mutant p53-R337H protein by calix[6]arene derivatives.

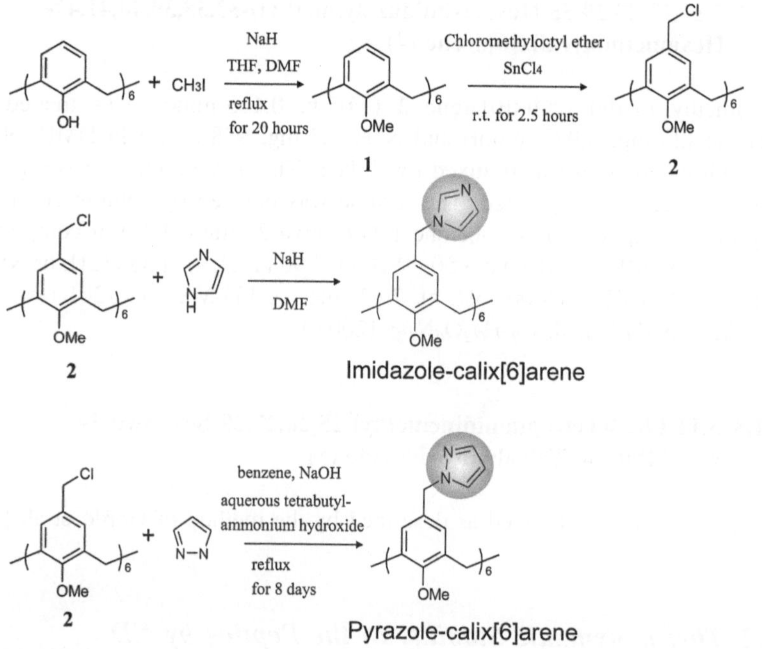

NaH
THF, DMF
reflux
for 20 hours

+ CH3I

Chloromethyloctyl ether
SnCl4
r.t. for 2.5 hours

OH   OMe   1   OMe   2   Cl

2

+

NaH
DMF

Imidazole-calix[6]arene

OMe

2

+

benzene, NaOH
aquerous tetrabutyl-
ammonium hydroxide
reflux
for 8 days

Pyrazole-calix[6]arene

OMe

Scheme 3.1 Synthesis of calixarene derivatives

3.2 Experimental Procedures

3.2.1 Synthesis of the Calixarene Derivatives

3.2.1.1 37,38,39,40,41,41-Hexamethoxycalix[6]arene (1)

Compound **1** was synthesized as described by the method of Gutsche and Lin [14] (Scheme 3.1).

3.2.1.2 5,11,17,23,29,35-Hexachloromethyl-37,38,39,40,41,41-Hexamethoxy-calix[6]arene (2)

Compound **2** was synthesized as described by the method of Almi et al. [15].

3.2.1.3 5,11,17,23,29,35-Hexa(1-pyrazolylmethyl)-37,38,39,40,-41,42-Hexamethoxycalix[6]arene (3)

Compound **3** was synthesized as described by the method of Zadmard et al. [16].

3.2.1.4 5,11,17,23,29,35-Hexa(1-imidazolylmethyl)-37,38,39,40,41,41-Hexamethoxycalix[6]arene (4)

Chloromethyl-methoxycalix[6]arene **2** (1.65 g, 0.163 mmol) was treated with imidazole (23 mg, 0.978 mmol) and NaH (23 mg, 978 μmol) in DMF (60 ml). The resulting solution was refluxed for 7 days. The organic phase was separated, washed, dried, and evaporated. The residue was purified by column chromatography over silica gel (CH_3Cl/acetone 1:1) to give 26 mg (11 %) of compound **4**. ^1H NMR (300 MHz, $CDCl_3$) δ 3.50 (18H, s), 3.90 (12H, s), 4.53 (12H, br. s), 6.69 (12H, s), 6.22 (6H, s), 7.00 (6H, s), 7.32 (6H, s). FD-MS m/z 1201 (M + H^+), 1223 (M + Na^+) [calcd. $C_{72}H_{72}O_6N_{12}$; 1200.6].

3.2.1.5 5,11,17,23-Tetraguanidinemethyl-25,26,27,28-biscrown-3-calix[4]arene Tetrahydrochloride (5)

Compound **5** was synthesized as described by the method of Gordo et al. [13].

3.2.2 Thermodynamic Stability of the Peptide by CD

CD spectra were recorded using a Jasco-805 spectropolarimeter in 50 mM sodium phosphate buffer containing 100 mM NaCl, pH 7.5. For thermal denaturation studies, spectra were recorded at discrete temperatures from 4 to 95 °C with a scan rate of 1 °C/min, and the ellipticity was measured at 222 nm for the p53TD solutions (10 μM monomer in 50 mM phosphate buffer, pH 7.5). The unfolding process of the p53TD peptide was fitted to a two-state transition model in which the native tetramer directly converts to an unfolded monomer. The thermodynamic parameters of the peptides were determined by calculation with the functions described by Mateu et al. [17]. We calculated the T_m and ΔH_u^{Tm} by fitting the fraction of monomer. Compound **3** was protonated by stirring with an excess of 2 M hydrochloric acid for 1 h. Compound 3-(6-HCl) was used.

3.2.3 Transcriptional Activity of p53-R337H in Cell

The transcriptional activity of the p53 R337H mutant in the presence of calixarene was measured by a modified method as previously described in Ref [4]. NCI-H1299 cells (p53-null mutant cell line) were cultured in a 35 mm dish in RPMI 1640 medium containing 10 % fetal calf serum. Cells were transfected with 0.5 μg of pEGFP-p53 and 3 μg of p53RE-mCherry reporter plasmid DNA with Lipofectamine 2000 (Invitrogen, USA) in OPTI-MEM. After 1 h, the medium was changed to RPMI-1640 medium with 10 % fetal calf serum and calixarene

Fig. 3.1 Structure of the p53 TD. Space-filling model of the p53 TD (PDB code 3sak) prepared with MolFeat version 4.0 (FiatLux Corp). Glutamic acid residues that interact with the calixarene derivatives (Glu336, Glu339 and Glu343) from two different monomers are shown in *blue*

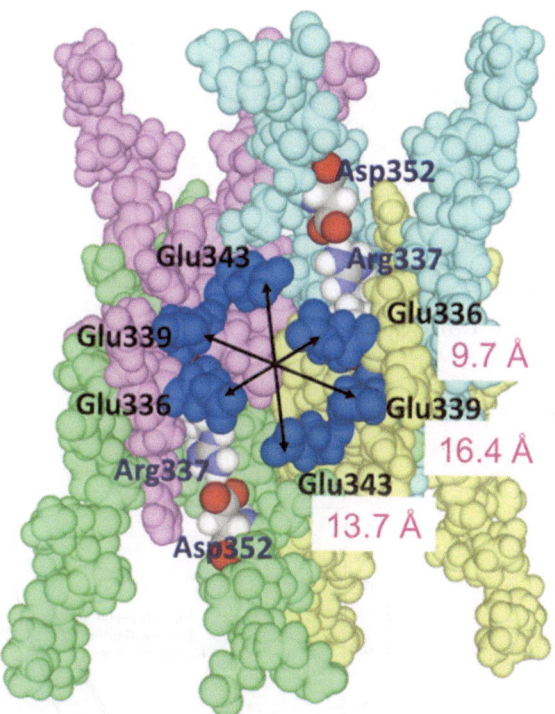

derivatives were added to the medium. After 11 h incubation, cells were fixed with 3.5 % formaldehyde, and EGFP (green) and mCherry (red) signals in the cells were quantified with a BZ-9000 (Keyence).

3.3 Results

3.3.1 Design of Calixarene Derivatives

On the tetrameric surface of p53, six glutamic acid residues (Glu346, Glu339 and Glu346) from two different monomers are located as shown in Fig. 3.1. In the tetrameric structure, distances between each glutamic acid residue are: Glu336-Glu336, 9.7 Å; Glu339-Glu339, 16.4 Å; Glu343-Glu343, 13.7 Å. We designed and synthesized two calix[6]arene derivatives that possess six imidazole (imidazole-calixarene **4**) [18] or six pyrazole (pyrazole-calixarene **3**) [16] groups at the upper rim to interact with the glutamic acid residues (Fig. 3.2, Scheme 3.1). The distance between each functional group of the calix[6]arenes is about 10–20 Å. The imidazole and pyrazole groups from the calix[6]arenes can interact with the glutamic acid residues and stabilize the mutant tetrameric structure. The guanidinium-calix[4]arene **5** was also synthesized, as described by Gordo et al. [13].

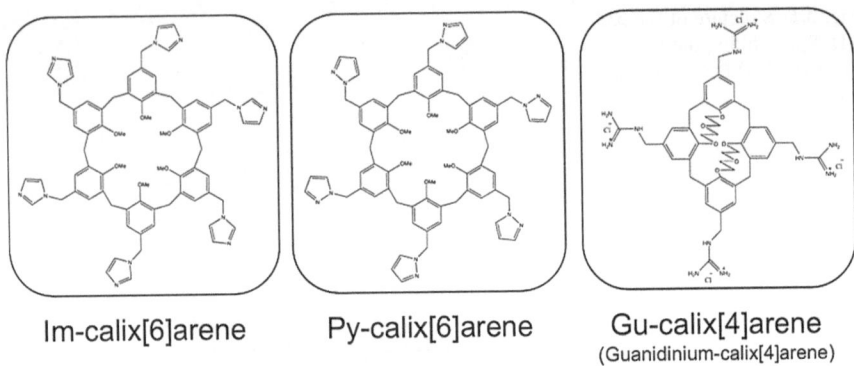

Im-calix[6]arene   Py-calix[6]arene   Gu-calix[4]arene
                                                    (Guanidinium-calix[4]arene)

Fig. 3.2 Structures of the calixarene derivatives

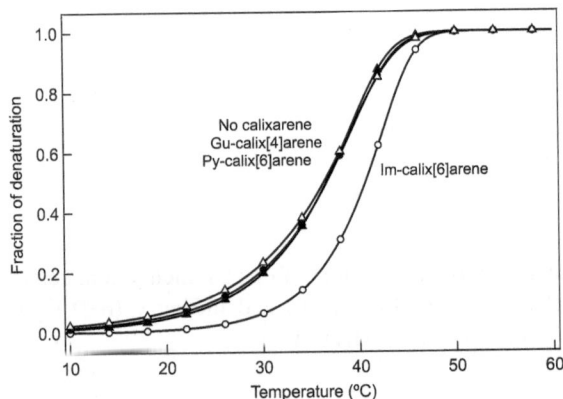

Fig. 3.3 Thermal denaturation curves of the p53TD-R337H peptide in the presence of calixarene derivatives. The unfolding process of the p53 TD peptide was fitted to a two-state transition model in which the native tetramer directly converts to an unfolded monomer. *Solid circle*, no derivative present; *open circle*, 40 μM imidazole-calixarene **4**; *solid triangle*, 10 μM pyrazole-calixarene **3**; *open triangle*, 40 μM guanidinium-calixarene **5**. Because pyrazole-calixarene **3** precipitated in phosphate buffer at the high concentrations used, the CD experiment was performed at 10 μM for pyrazole-calixarene **3**

3.3.2 Stabilization of Mutant Peptides by Calixarene Derivatives

Mutant p53TD peptide (R337H, E339K, and E343G) corresponding to p53 TD residues 319–358, was chemically synthesized as described previously [19]. The thermal denaturation curves of the mutant p53TD peptide (10 μM monomer) in the presence of calixarenes were calculated from changes in CD ellipticity at 222 nm, which corresponds to an α-helix (Fig. 3.3, Table 3.1). The results showed that

Table 3.1 Thermodynamic parameters for the p53TD peptide in the presence of calixarenes

Mutants	Calixarenes	T_m (°C)	ΔT_m (°C)	ΔH_u^{Tm} (kcal/mol)	$\Delta\Delta G_u^{37\,°C}$ (kcal/mol)
R337H	No calixarene	36.6 ± 0.1	–	116.9 ± 2.4	–
	Imidazole-calixarene **1**	40.7 ± 0.2	4.1	167.6 ± 5.4	−2.06
	Pyrazolea-calixarene **2**	36.6 ± 0.2	0.0	124.2 ± 3.4	0.04
	Guanidium-calixarene **3**	36.3 ± 0.2	−0.3	108.2 ± 4.3	0.10
R337L	No calixarene	36.2 ± 0.2	–	75.4 ± 2.9	–
	Imidazole-calixarene **1**	39.8 ± 0.2	3.6	87.9 ± 2.7	−0.97
	Pyrazole-calixarene **2**	38.6 ± 0.3	2.4	95.9 ± 4.4	−0.67
	Guanidium-calixarene **3**	38.5 ± 0.3	2.3	66.0 ± 4.3	−0.50
WT	No calixarene	68.7 ± 0.2	–	154.9 ± 3.7	–
	Imidazole-calixarene **1**	68.9 ± 0.1	0.2	126.6 ± 2.4	2.59
	Pyrazole-calixarene **2**	68.7 ± 0.1	0.0	149.1 ± 4.4	0.29
	Guanidium-calixarene **3**	67.7 ± 0.2	−1.0	143.4 ± 2.0	0.78
E339K	No calixarene	70.0 ± 0.2	–	153.8 ± 5.0	–
	Imidazole-calixarene **1**	67.3 ± 0.1	−2.7	121.7 ± 3.2	3.55
	Pyrazole-calixarene **2**	70.7 ± 0.2	0.7	145.6 ± 6.4	0.64
	Guanidium-calixarene **3**	69.0 ± 0.2	−1.0	151.9 ± 4.8	0.42
E343G	No calixarene	61.9 ± 0.2	–	151.9 ± 4.9	–
	Imidazole-calixarene **1**	57.7 ± 0.2	−4.2	143.0 ± 4.2	1.83
	Pyrazole-calixarene **2**	58.2 ± 0.2	−3.7	130.4 ± 4.0	2.49
	Guanidium-calixarene **3**	62.2 ± 0.2	0.3	116.5 ± 4.5	2.57

T_m, transition temperature; ΔH_u^{Tm}, variation in the enthalpy of unfolding at T_m; $\Delta\Delta G_u^{37\,°C}$, the difference in ΔG between p53TD peptide in the absence of calixarene derivatives and the peptide in the presence of calixarene derivatives at 37 °C. The standard errors of fittings are indicated

imidazole-calixarene **4** (40 μM) significantly stabilized tetramer formation by R337H, with the T_m value shifted from 36.6 to 40.7 °C. The $\Delta\Delta G_u$ value calculated at 37 °C was $\Delta\Delta G_u^{37\,°C} = -2.06$ kcal/mol. It is reported that guanidinium-calix[4]arene **5** stabilized R337H tetrameric structure in the absence of salt (Fig. 3.4) [13]. However, compound **5** showed no stabilization for R337H in the presence of 100 mM NaCl. For pyrazole-calixarene **3** (10 μM), the stability of the p53TD-R337H peptide was not significantly altered ($\Delta\Delta G_u^{37\,°C} = 0.04$ kcal/mol.). This difference in stabilization could be explained by the pK_a values of the imidazole and pyrazole groups. The pK_a of a 1-methylimidazole is 7.06 [20], so half of the imidazole groups of the calix[6]arene derivative could be protonated under the conditions used in the CD experiment. By contrast, the pyrazole group would be mainly unprotonated, because the pK_a of a 1-methylpyrazole is 2.04 [21]. The data suggested that imidazole-calixarene **4** stabilized the tetrameric structure of R337H under physiological conditions, possibly through interactions with the glutamic acid residues.

For the wild-type p53TD peptide, all calixarene derivatives did not enhance the stability of the tetrameric structure (Fig. 3.5, Table 3.1). This result indicated that calixarene derivatives could not bind the tetrameric structure of p53TD at high

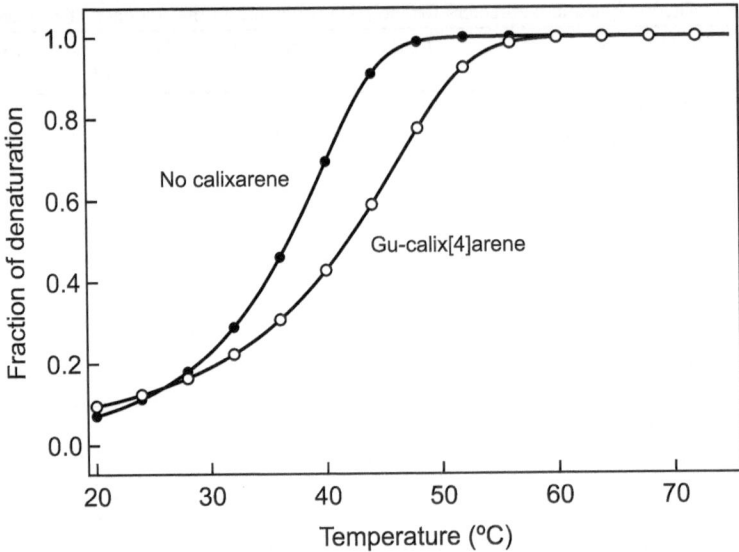

Fig. 3.4 Thermal denaturation curves of the p53TD R337H peptide in the presence of guanidinium-calix[4]arene **5**. The unfolding process of the p53 TD peptide was fitted to a two-state transition model in which the native tetramer directly converts to an unfolded monomer. *Solid circle*, no compound present; *open circle*, 20 μM guanidinium-calixarene **5**. Samples were prepared in water and the pH was thoroughly adjusted to 7.0

temperature such as the melting temperature of wild-type p53TD peptide because of the low binding affinity of derivatives for p53TD. Furthermore, the calixarene derivatives could not stabilized Glu residue mutants (E339K and E343G), and moreover, these mutants were destabilized by the calixarene derivatives (Figs. 3.6, 3.7, Table 3.1). Mutation of the solvent-exposed residues changed the electrostatic potential on the surface of the p53TD, including potential interaction site of the calixarene derivatives (Figs. 3.6, 3.7). We suggested that surface charge change might result into charge/charge repulsion between the calixarene derivatives and p53TD. The data suggested that imidazole-calixarene 1 stabilized the tetrameric structure of the mutant peptide under physiological conditions, possibly through interactions with the glutamic acid residues.

3.3.3 Transcriptional Activity of p53 Protein in Cell

The imidazole-calixarene **4** stabilized the mutant tetrameric structure of R337H in in vitro assays. We analyzed the effects of imidazole-calixarene **1** on the transcriptional activity of the p53-R337H protein in cells. The transcriptional activity of the p53-R337H mutant in the presence of calixarene was measured by a modified method as previously described [4, 22]. H1299 cells were transfected with the EGFP-p53-R337H expression vector and the reporter plasmid p53RE-mCherry(NLS) (Fig. 3.8).

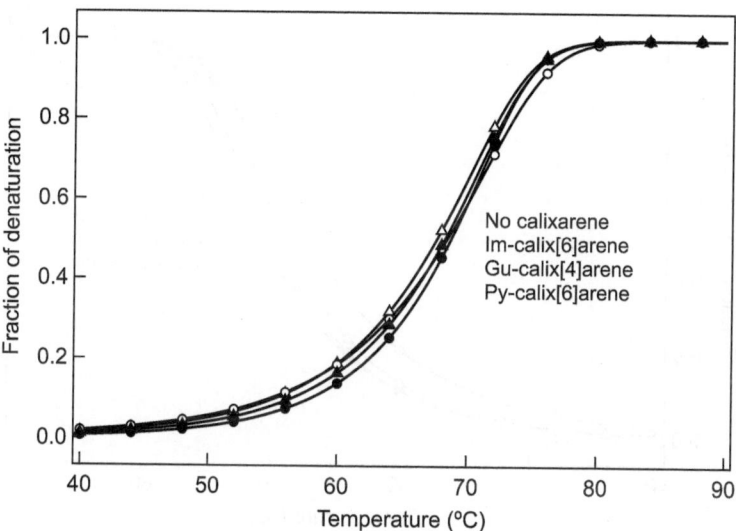

Fig. 3.5 Thermal denaturation curves of the p53TD-WT. The unfolding process of the p53 TD peptide was fitted to a two-state transition model in which the native tetramer directly converts to an unfolded monomer. *Solid circle,* no derivative present; *open circle,* 40 μM imidazole-calixarene **4**; *solid triangle,* 10 μM pyrazole-calixarene **3**; *open triangle,* 40 μM guanidinium-calixarene **5**. Because pyrazole-calixarene **3** precipitated in phosphate buffer at the high concentrations used, the CD experiment was performed at 10 μM for pyrazole-calixarene **3**

In this study, we used p53RE-mCherry(NLS) plasmid as the reporter vector, instead of p53RE-DsRed, to enhance the red fluorescence signal. The amount of p53 expressed in each cell and its transcriptional activity were estimated from the green and red fluorescence signals, respectively. Cells expressing EGFP-p53-R337H mutant in the absence of calixarene derivatives showed weak mCherry signals (Fig. 3.9, left panels). The mCherry signal in each cell increased in the presence of imidazole-calixarene **1** (Fig. 3.9, right panels). The red signal in each cell was quantified, which expressed EGFP-p53 at the same level as endogenous p53 in A549, estimated by immunostain with anti-p53 monoclonal antibody [4]. Many human normal cells and cancer cell lines with the wild-type p53 are reported to express p53 protein mostly in the same range [23, 24]. Figure 3.10 shows that the transcriptional activity of EGFP-p53-R337H increased to ~ 130 % in the presence of imidazole-calixarene **4** compared with the absence of the compound. On the other hand, guanidinium-calixarene **5**, which showed no stabilization of the mutant tetrameric structure, did not enhance the transcriptional activity of EGFP-p53-R337H in cells (Fig. 3.10). For the wild-type p53 or an inactive mutant p53 with the mutation of R273H in DNA binding domain, imidazole-calixarene **4** did not show any enhancement of the transcriptional activities (Fig. 3.11). These results suggested that imidazole-calixarene **4** could restore the transcriptional activity of p53-R337H through stabilization of its tetrameric structure.

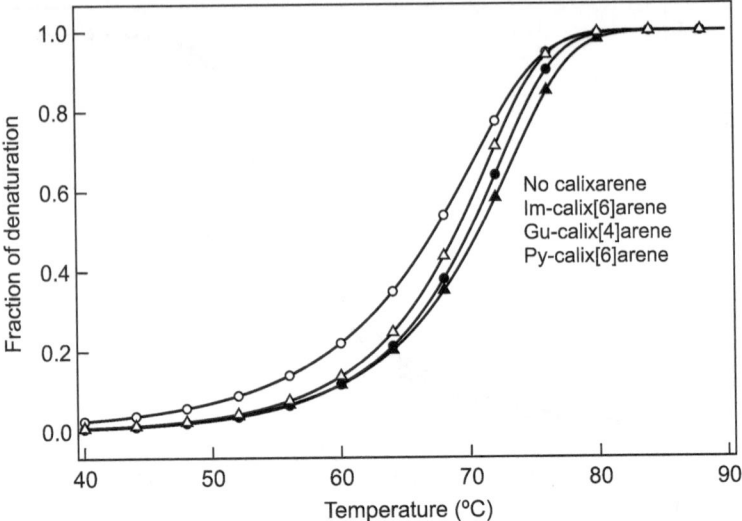

Fig. 3.6 Thermal denaturation curves of the p53TD-E339K peptides. The unfolding process of the p53 TD peptide was fitted to a two-state transition model in which the native tetramer directly converts to an unfolded monomer. *Black*, no derivative present; *open circle*, 40 μM imidazole-calixarene **4**; *solid triangle*, 10 μM pyrazole-calixarene **3**; *open triangle*, 40 μM guanidinium-calixarene **5**. Because pyrazole-calixarene **3** precipitated in phosphate buffer at the high concentrations used, the CD experiment was performed at 10 μM for pyrazole-calixarene **3**

3.4 Discussion

The calix[6]arene derivative with six imidazole groups could stabilize the p53 tetrameric structure under physiological conditions. In the absence of salt (in H_2O at pH 7.0), guanidinium-calix[4]arene **5**, which has four guanidinium groups, could induce significant thermal stabilization in mutant R337H ($\Delta T_m = \sim 10$ °C, 20 μM of compound **5** for 10 μM of R337H) [13]. However, guanidinium-calix[4]arene **5** does not show stabilization of the mutant p53 under physiological conditions. These data indicate one advantage of using calix[6]arene as a template in restoring mutant p53 function. Furthermore, we have demonstrated that imidazole-calixarene **4** could enhance the transcriptional activity of p53-R337H in cells. The future optimization of this calix[6]arene derivative might lead to the development of a candidate for cancer therapy.

In Chap. 2, we demonstrated that relatively small destabilization of the tetrameric structure by the missense mutation could induce significantly decrease of the tetramer in nucleus. In this chapter, I showed the possibility that small stabilization by imidazole-calix[6]arene **4** could enhance the transcriptional activity of p53. This result might supports the hypothesis that destabilization (or stabilization) of the tetrameric structure, which is deemed to be relatively small change in vitro, is amplified and could cause decrease (or increase) the transcriptional activity of p53

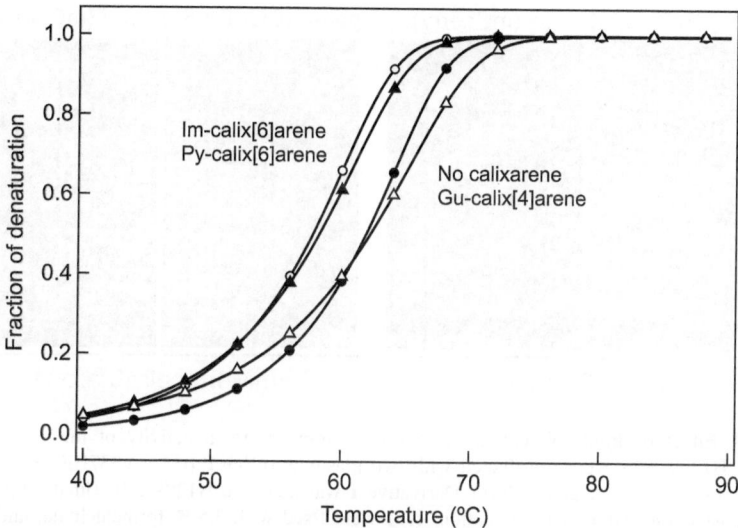

Fig. 3.7 Thermal denaturation curves of the p53TD-E343G peptides. The unfolding process of the p53 TD peptide was fitted to a two-state transition model in which the native tetramer directly converts to an unfolded monomer. *Black*, no derivative present; *open circle*, 40 μM imidazole-calixarene **4**; *solid triangle*, 10 μM pyrazole-calixarene **3**; *open triangle*, 40 μM guanidinium-calixarene **5**. Because pyrazole-calixarene **3** precipitated in phosphate buffer at the high concentrations used, the CD experiment was performed at 10 μM for pyrazole-calixarene **3**

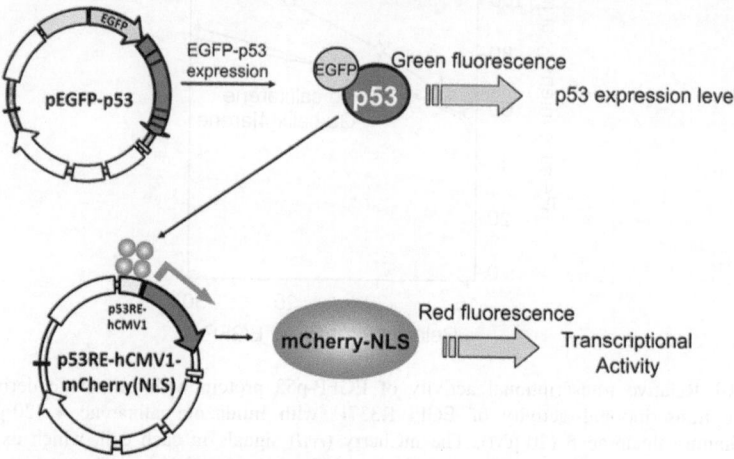

Fig. 3.8 Analysis of the transcriptional activity of p53 protein. The expression of mCherry(NLS) was dependent on the transcriptional activity of EGFP-p53. The amount of EGFP-p53 expressed in each cell and its transcriptional activity were estimated from the green and red fluorescence signals, respectively

green (EGFP-p53)   red (mCherry)   green (EGFP-p53)   red (mCherry)

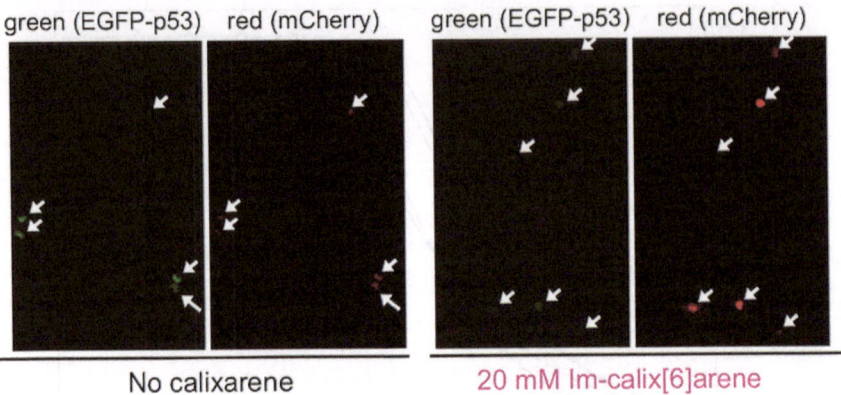

No calixarene   20 mM Im-calix[6]arene

Fig. 3.9 Effect of imidazole-calixarene **4** on the transcriptional activity of the p53 R337H mutant. H1299 cells were transfected with two plasmids (pEGFP-p53-Arg337His and p53RE-mCherry) using Lipofectamine 2000. Derivative **1** was added to H1299 cells (final 20 μM) 1 h after transfection. After 11 h incubation, cells were fixed with 3.5 % formaldehyde, and green and red fluorescence signals (*arrows*) were quantified in each cell

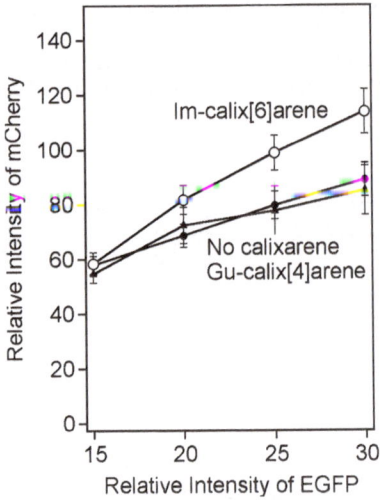

Fig. 3.10 Relative transcriptional activity of EGFP-p53 protein with calixarene derivatives. Relative transcriptional activity of EGFP-R337H with imidazole-calixarene **4** (20 μM) or guanidinium-calixarene **5** (20 μM). The mCherry (*red*) signals in each cell, which expressed EGFP-p53 at same level as endogenous p53 in A549 were quantified [4]. The averages of the mCherry signals in each EGFP signal are shown. *Solid circle*, no derivative present; *open circle*, 20 μM imidazole-calixarene **1**; *solid triangle*, 20 μM guanidinium-calixarene **5**. The standard error is indicated

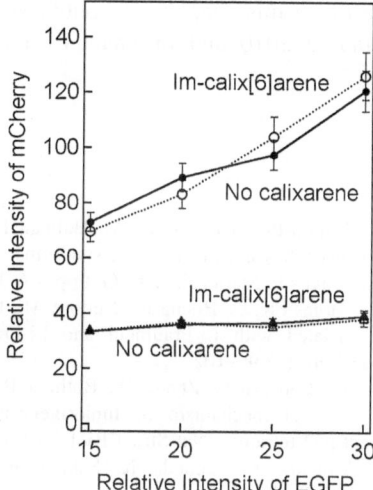

Fig. 3.11 Relative transcriptional activity of the wild-type EGFP-p53 and EGFP-p53-R273H with imidazole-calixarene **4** (20 μM). The averages of the mCherry signals in each EGFP signal are shown. *Solid circle solid line*, wild-type EGFP-p53 without compound **4**; *open circle dotted line*, wild-type EGFP-p53 with compound **4**; *solid triangle solid line*, R273H without compound **4**; *open triangle dotted line*, R273H with compound **4**. The standard error is indicated. The data were averaged from three independent experiments

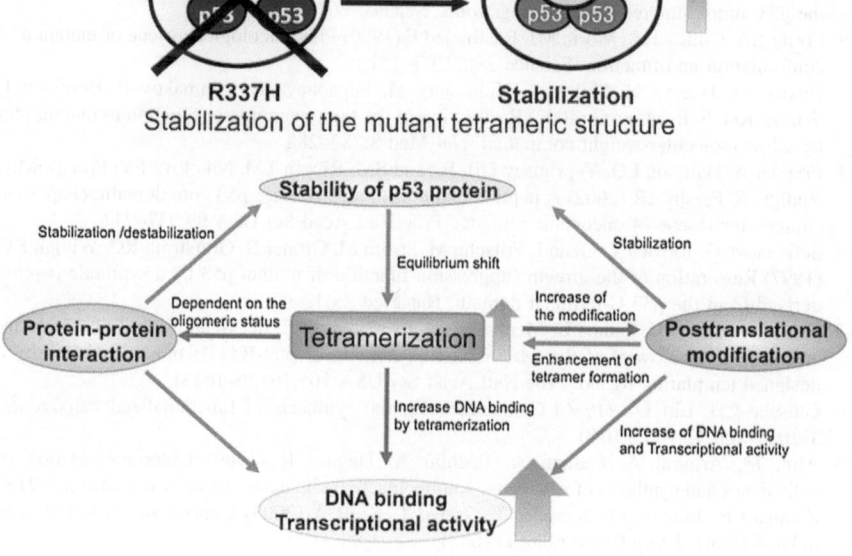

Fig. 3.12 Stabilization of mutant tetrameric structure by calixarene derivatives

(Fig. 3.12). The main result in this chapter was published first in *Bioorganic & Medicinal Chemistry Letters* in 2010, and this chapter is its expanded version [25].

References

1. Palmero EI, Achatz MI, Ashton-Prolla P, Olivier M, Hainaut P (2010) Tumor protein 53 mutations and inherited cancer: beyond Li-Fraumeni syndrome. Curr Opin Oncol 22:64–69
2. Achatz MI, Olivier M, Le Calvez F, Martel-Planche G, Lopes A, Rossi BM, Ashton-Prolla P, Giugliani R, Palmero EI, Vargas FR, Da Rocha JC, Vettore AL, Hainaut P (2007) The TP53 mutation, R337H, is associated with Li-Fraumeni and Li-Fraumeni-like syndromes in Brazilian families. Cancer Lett 245:96–102
3. DiGiammarino EL, Lee AS, Cadwell C, Zhang W, Bothner B, Ribeiro RC, Zambetti G, Kriwacki RW (2002) A novel mechanism of tumorigenesis involving pH-dependent destabilization of a mutant p53 tetramer. Nat Struct Biol 9:12–16
4. Imagawa T, Terai T, Yamada Y, Kamada R, Sakaguchi K (2009) Evaluation of transcriptional activity of p53 in individual living mammalian cells. Anal Biochem 387:249–256
5. Kawaguchi T, Kato S, Otsuka K, Watanabe G, Kumabe T, Tominaga T, Yoshimoto T, Ishioka C (2005) The relationship among p53 oligomer formation, structure and transcriptional activity using a comprehensive missense mutation library. Oncogene 24:6976–6981
6. Sakamoto H, Lewis MS, Kodama H, Appella E, Sakaguchi K (1994) Specific sequences from the carboxyl terminus of human p53 gene product form anti-parallel tetramers in solution. Proc Natl Acad Sci USA 91:8974–8978
7. Clore GM, Ernst J, Clubb R, Omichinski JG, Kennedy WMP, Sakaguchi K, Appella E, Gronenborn AM (1995) Refined solution structure of the oligomerization domain of the tumour suppressor p53. Nat Struct Biol 2:321–333
8. Jeffrey PD, Gorina S, Pavletich NP (1995) Crystal structure of the tetramerization domain of the p53 tumor suppressor at 1.7 angstroms. Science 267:1498–1502
9. Foster BA, Coffey HA, Morin MJ, Rastinejad F (1999) Pharmacological rescue of mutant p53 conformation and function. Science 286:2507–2510
10. Bykov VJ, Issaeva N, Shilov A, Hultcrantz M, Pugacheva E, Chumakov P, Bergman J, Wiman KG, Selivanova G (2002) Restoration of the tumor suppressor function to mutant p53 by a low-molecular-weight compound. Nat Med 8:282–288
11. Friedler A, Hansson LO, Veprintsev DB, Freund SM, Rippin TM, Nikolova PV, Proctor MR, Rudiger S, Fersht AR (2002) A peptide that binds and stabilizes p53 core domain: chaperone strategy for rescue of oncogenic mutants. Proc Natl Acad Sci USA 99:937–942
12. Selivanova G, Iotsova V, Okan I, Fritsche M, Strom M, Groner B, Grafstrom RC, Wiman KG (1997) Restoration of the growth suppression function of mutant p53 by a synthetic peptide derived from the p53 C-terminal domain. Nat Med 3:632–638
13. Gordo S, Martos V, Santos E, Menendez M, Bo C, Giralt E, de Mendoza J (2008) Stability and structural recovery of the tetramerization domain of p53-R337H mutant induced by a designed templating ligand. Proc Natl Acad Sci USA 105:16426–16431
14. Gutsche CD, Lin LG (1986) Calixarenes 12: the synthesis of functionalized calixarenes. Tetrahedron 42:1633–1640
15. Almi M, Arduini A, Casnati A, Pochini A, Ungaro R (1989) Chloromethylation of calixarenes and synthesis of new water soluble macrocyclic hosts. Tetrahedron 45:2177–2182
16. Zadmard R, Junkers M, Schrader T, Grawe T, Kraft A (2003) Capsule-like assemblies in polar solvents. J Org Chem 68:6511–6521

17. Mateu MG, Fersht AR (1998) Nine hydrophobic side chains are key determinants of the thermodynamic stability and oligomerization status of tumour suppressor p53 tetramerization domain. EMBO J 17:2748–2758

18. Atz J, Wagner P, Roemer K (2000) Function, oligomerization, and conformation of tumor-associated p53 proteins with mutated C-terminus. J Cell Biochem 76:572–584

19. Nomura T, Kamada R, Ito I, Chuman Y, Shimohigashi Y, Sakaguchi K (2009) Oxidation of methionine residue at hydrophobic core destabilizes p53 tetrameric structure. Biopolymers 91:78–84

20. Casciola M, Costantino U, Calevi A (1993) Intercalation compounds of zirconium phosphates with substituted pyrazoles and imidazoles and their ac conductivity. Solid State Ionics 61:245–250

21. Boulton BE, Coller BAW (1971) Kinetics, stoicheiometry, and mechanism in the bromination of aromatic heterocycles. I. Aust J Chem 24:1413–1423

22. Bergamaschi D, Samuels Y, Sullivan A, Zvelebil M, Breyssens H, Bisso A, Del Sal G, Syed N, Smith P, Gasco M, Crook T, Lu X (2006) iASPP preferentially binds p53 proline-rich region and modulates apoptotic function of codon 72-polymorphic p53. Nat Genet 38:1133–1141

23. Wang YV, Wade M, Wong E, Li YC, Rodewald LW, Wahl GM (2007) Quantitative analyses reveal the importance of regulated Hdmx degradation for p53 activation. Proc Natl Acad Sci USA 104:12365–12370

24. Chen L, Lu W, Agrawal S, Zhou W, Zhang R, Chen J (1999) Ubiquitous induction of p53 in tumor cells by antisense inhibition of MDM2 expression. Mol Med 5:21–34

25. Kamada R, Yoshino W, Nomura T, Chuman Y, Imagawa T, Suzuki T, Sakaguchi K (2010) Enhancement of transcriptional activity of mutant p53 tumor suppressor protein through stabilization of tetramer formation by calix[6]arene derivatives. Bioorg Med Chem Lett 20:4412–4415

Chapter 4
Inhibition of the Transcriptional Activity of p53 Through Hetero-Oligomerization

4.1 Introduction

Induced pluripotent stem (iPS) cells can be generated from skin fibroblasts by expressing pluripotency factors and oncogenes (*KLF4*, *SOX2*, *OCT4*, and *c-MYC*) [1]. Three to four weeks after viral vector-mediated transfection of these pluripotency factors, small numbers of transfected cells begin to become morphologically and biochemically similar to pluripotent stem cells. iPS cells can be differentiated into various cell types, in a similar manner to ES cells. However, the viral transfection systems used result in the random integration of the pluripotency genes in the host's genome. This raises issues for the potential therapeutic application of iPS cells because such random retroviral integration might increase the risk of tumor formation [2]. To overcome these dangers, reprogramming was accomplished using a plasmid-based system, although the efficiency of reprogramming adult fibroblasts was low (<0.1 %) [3]. Recently, it was reported that over-expression of reprogramming factors, including the oncogene c-Myc, can activate the p53-p21 pathway [4], which reduces reprogramming efficiency. Thus, reducing p53 activity by expressing a dominant-negative p53 mutant or by deleting or knocking down p53 or p21 increases reprogramming efficiency (Fig. 4.1). Notably, however, suppression of p53 may increase the risk of cancer development. p53-deficient mice are extremely sensitive to radiation-induced tumorigenesis. Thus, temporary suppression of p53 is necessary for the efficient generation of iPS cells.

The tumor suppressor protein p53 acts as a homotetramer and tetramer formation of p53 is essential for its tumor suppressor function. p53 induces transcription of cell cycle arrest- and apoptosis-inducing genes, such as *p21* and *bax*. Approximately 50 % of human tumors carry inactivating mutations in the p53 gene. p53 DNA binding domain mutants, which have lost their DNA binding ability, inactivate wild-type p53 by hetero-oligomerization through the tetramerization domains. The p53 tetramerization domain consists of a β-strand (Glu326-Arg333), a tight turn (Gly334), and an α-helix (Arg335-Gly356). Two monomers

R. Kamada, *Tetramer Stability and Functional Regulation of Tumor Suppressor Protein p53*, Springer Theses, DOI: 10.1007/978-4-431-54135-6_4, © Springer Japan 2012

Fig. 4.1 Generation of induced pluripotent stem cells (iPS cells). Reprogramming efficiency is reduced through oncogene-mediated activation of the p53 pathway. Thus, reduced signaling of p53 by expressing a dominant-negative mutant p53 or by deleting or knocking down p53 or p21 increases reprogramming efficiency

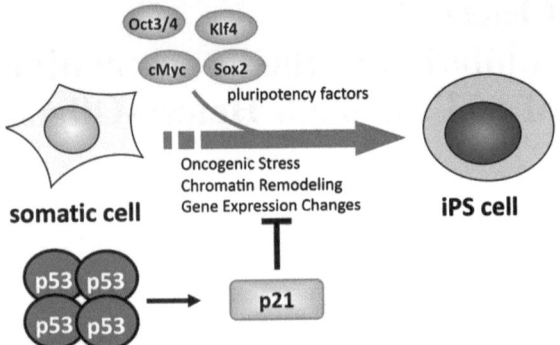

form a dimer through interaction between antiparallel β-sheets and α-helices, and the two dimers associate to form a tetramer with an unusual four-helix bundle. The p53 tetramerization domain peptide, in the absence of other domains, including the DNA binding domain, might bind to full-length p53 and inhibit p53's function. To test this hypothesis, the tetramerization domain peptide was introduced into cells.

The plasma membrane of the cell forms an effective barrier, which restricts the intracellular uptake of macromolecules. Several proteins (containing a protein transduction domain or PTD), such as the HIV-1 transcriptional activator Tat protein and the herpes simplex virus structural protein, VP22, have been discovered which have the ability to efficiently pass through the plasma membrane of eukaryotic cells [5]. Not only can these proteins pass through the plasma membrane but they can enable other proteins or peptides that are attached to be introduced into cells. The transduction of these proteins does not appear to be affected by cell type and they can be efficiently introduced into cells both in vitro and in vivo with no apparent toxicity. The efficiency of cellular uptake of the transduction domain strongly correlates with the number of basic residues present, indicating that internalization is dependent on an interaction between the charged side groups of the basic residues and lipid phosphates on the cell surface. Homopolymers of L-arginine or L-lysine can also show efficient internalization.

In this study, tetramerization domain peptides fused with a PTD were synthesized to introduce the peptides into cells. The PTD-p53tet peptide could be efficiently introduced into cells and it significantly decreased the transcriptional activity of p53.

4.2 Experimental Procedures

4.2.1 Peptide Synthesis

PTDs attached to p53 tetramerization domain peptides were synthesized using standard Fmoc chemistry. To visualize the subcellular localization of PTD-p53tet peptides, K10-p53tet peptide was labeled with fluorescein at the N-terminus. K10-

p53tet peptide resin was swollen in DMF, washed with NMP and then incubated with 5-(and 6-)carboxyfluorescein, succinimidyl ester (Thermo scientific Inc., Bremen, Germany) in NMP for 6 h. Peptide resin cleavage and peptide deprotection were accomplished in a single step using regent K (TFA:H_2O:thioanisole:ethanedithiol:phenol = 82.5:5:5:2.5:5) to yield a crude peptide amide. All synthetic peptides were purified by reverse-phase HPLC on a Vydac C-8 column with elution in a linear gradient of water containing 0.05 % TFA and acetonitrile containing 0.04 % TFA. The correct sequence of purified peptides was confirmed by matrix-assisted laser desorption/ionization time-of-flight mass spectrometry (MALDI-TOF–MS).

4.2.2 CD Analysis of PTD-p53tet Peptide

CD spectra were recorded on a JASCO J-805 spectropolarimeter (JASCO, Tokyo, Japan) using a 1 mm path length quartz cell. A Refrigerated/Heating Circulator (Julabo, Seelbach, Germany) was used to control the temperature of the cell. The peptide concentrations were 10 µM in 50 mM sodium phosphate buffer, pH 7.5, containing 100 mM NaCl. CD spectra were the average of six measurements obtained by collecting data from 260 to 195 nm at a rate of 50 nm/min at 37 °C.

4.2.3 Hetero-Oligomerization of PTD-p53tet Peptide

K10- or R10-p53tet peptides were mixed with an equal concentration of biotinylated p53(319-358) in 50 mM sodium phosphate buffer, pH 7.5, containing 100 mM NaCl. The peptide mixtures were incubated with Avidin beads (Thermo scientific Inc., Bremen, Germany) for 3 h at 4 °C and the beads were then washed with 50 mM sodium phosphate buffer. The peptides were extracted with 80 % acetic acid and analyzed by analytical HPLC on a C-8 reverse-phase column.

4.2.4 Transcriptional Activity of p53 with PTD-p53tet Peptide

The transcriptional activity of wild-type p53 in the presence of PTD-p53tet peptide was measured using a modification of the method previously described in Ref. [2]. NCI-H1299 cells (a p53-null mutant cell line) were cultured in a 35 mm dish in RPMI 1640 medium containing 10 % fetal calf serum. Six hours before transfection, peptides were added to the cells. Cells were transfected with 0.8 µg of p53RE-mCherry-2-Venus-p53 with Lipofectamine 2000 (Invitrogen, USA) in OPTI-MEM. After 1 h, RPMI 1640 medium with 10 % FBS was added and cells were cultured. After 11 h, the cells were fixed with 3.5 % formaldehyde, stained with DAPI, and

Table 4.1 Amino acid sequences of the PTD-p53tet peptides

K10-p53tet

H-KKKKKKKKKK [p53(322-358)-NH₂]

K10-p53tet-L344P (mutant)

H-KKKKKKKKKK [p53(322-358)-L344P-NH₂]

K10-p53tet

H-RRRRRRRRRR [p53(322-358)-NH₂]

K10-p53tet-L344P (mutant)

H-RRRRRRRRRR [p53(322-358)-L344P-NH₂]

Fluor-K10-p53tet

Fluorescein-KKKKKKKKKK [p53(322-358)-NH₂]

EGFP (green) and mCherry (red) signals in the cells were quantified with a BZ-9000 fluorescence microscope (Keyence Corp., Osaka, Japan).

4.3 Results

4.3.1 CD Spectra of PTD-p53tet Peptides

The objective of this study was to inhibit p53 transcriptional activity by hetero-oligomerization of the p53 tetramerization domain peptide with full-length p53. To introduce the tetramerization domain peptide into cells, PTDs attached to tetramerization domain peptides (PTD-p53tet) were designed. For PTD sequence, polyarginine and polylysine sequences were used and four PTD-p53tet peptides, K10-p53tet-WT, K10-p53tet-L344P, R10-p53tet-WT, and R10-p53tet-L344P, were chemically synthesized by standard Fmoc methodology (Table 4.1). The secondary structure of PTD-p53tet peptides was analyzed by CD. K10-p53tet-WT and R10-p53tet-WT showed negative minima at 208 and 222 nm, which is characteristic of the spectrum for the p53 tetramer structure (Fig. 4.2). This result suggested that polylysine and polyarginine sequences fused to the N-terminus did not interrupt tetramer formation of p53. K10- and R10-p53tet-L344P peptides existed as a random coil. These results suggested that K10- and R10-p53tet-WT peptides could form a wild-type p53-like tetrameric structure. K10- and R10-p53tet-L344P peptides, which have a Pro→Leu mutation in the α-helix, could not form a tetramer and existed as random-coil monomers.

4.3.2 Hetero-Oligomerization of PTD-p53tet Peptides with p53(319-358) Peptide

We investigated whether the PTD-p53tet peptides could form a hetero-oligomer with the p53(319-358) wild-type peptide. K10- and R10-p53tet-WT peptides, but

Fig. 4.2 CD spectra of PTD-p53tet peptides. CD spectra of K10- and R10-p53tet peptides in 50 mM phosphate buffer, pH 7.5, 100 mM NaCl at 37°C. Peptide concentration was 10 μM. The CD spectra were recorded on a Jasco-805 s (JASCO) spectropolarimeter using a 1 mm path length quartz cell. *Solid circle solid line* p53(319-358); *open circle solid line* K10-p53tet-WT; *solid triangle solid line* K10-p53tet-L344P; *open triangle dotted line* R10-p53tet-WT; *solid inversed triangle dotted line* R10-p53tet-L344P

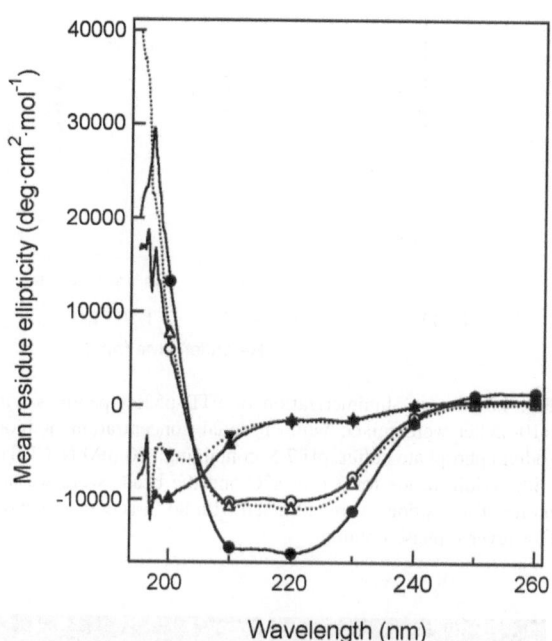

not L344P mutants, were coprecipitated with the biotinylated p53(319-358) peptide (Fig. 4.3). This result indicated that the PTD-p53tet-WT peptides could form hetero-oligomers with the wild-type tetramerization domain sequence. This suggested that the PTD sequence, (K10 and R10) did not interfere with tetramerization of the peptide.

4.3.3 Cellular Introducing and Localization of PTD-p53tet Peptides

To visualize the introduction and the subcellular localization of PTD-p53tet peptide, H1299 cells were exposed to fluorescein-labeled K10-p53tet-WT peptide for 6 h. The H1299 cells were then stained with DAPI and observed using a BZ9000 microscope (Keyence Corp., Osaka, Japan). The Fluor-K10-p53tet-WT peptide was observed in cells and was localized in the nucleus (Fig. 4.4).

4.3.4 Transcriptional Activity of Venus-p53 Protein in Cell

The effect of PTD-p53tet peptides on the transcriptional activity of p53 was analyzed using a p53 reporter system employing yellow fluorescent protein for p53

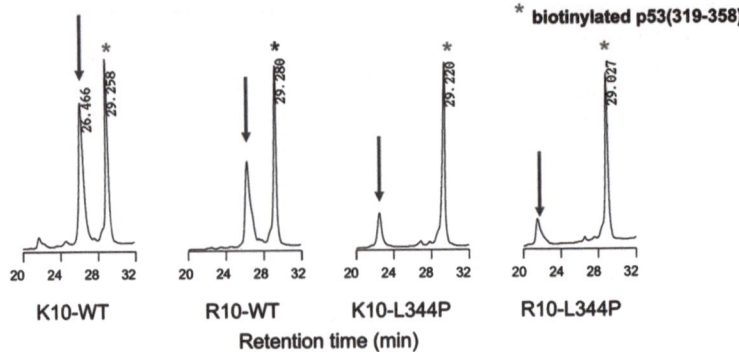

Fig. 4.3 Hetero-oligomerization of PTD-p53tet peptides with p53(319-358) peptide. K10- or R10-p53tet were mixed with an equal concentration of biotinylated p53(319-358) in 50 mM sodium phosphate buffer, pH 7.5, containing 100 mM NaCl. The peptide mixtures were incubated with Avidin beads for 3 h at 4°C and the beads were washed with 50 mM sodium phosphate buffer. The peptides were extracted with 80 % acetic acid and analyzed by analytical HPLC on a C-8 reverse-phase column

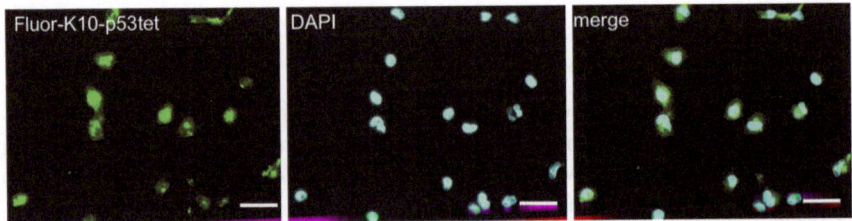

Fig. 4.4 Introduction and cellular localization of Fluor-K10-p53tet peptide. H1299 cells were exposed to Fluor-K10-p53tet (10 μM) peptide for 24 h. The cells were fixed with 3.5 % formaldehyde, treated with 0.2 % Triton X-100 for 5 min at room temperature and then washed three times with PBS. The cells were stained with DAPI. *Scale bars* indicate 50 μm

protein levels and red fluorescent protein for transactivation by p53. This system contains a Venus-p53 expression unit and an mCherry(NLS) expression unit with a p53-dependent promoter in a single plasmid (Fig. 4.5). The ratio between Venus-p53 expression and mCherry reporter expression is constant; therefore, transcriptional activity of p53 can be accurately measured. In the absence of the PTD-p53tet peptide, cells expressing Venus-p53 showed strong mCherry signals (Fig. 4.5). In the presence of K10- and R10-p53tet-WT peptides, which can form hetero-oligomers with wild-type p53(319-358), mCherry signals in cells were significantly decreased (Fig. 4.6). In contrast, mutant K10- and R10-p53tet peptides, which could not bind to wild-type p53(319-358) peptide, did not affect mCherry signals. The mCherry signals in each cell, which expressed EGFP-p53 at the same level as endogenous p53 in A549 cells, were quantified. As shown in

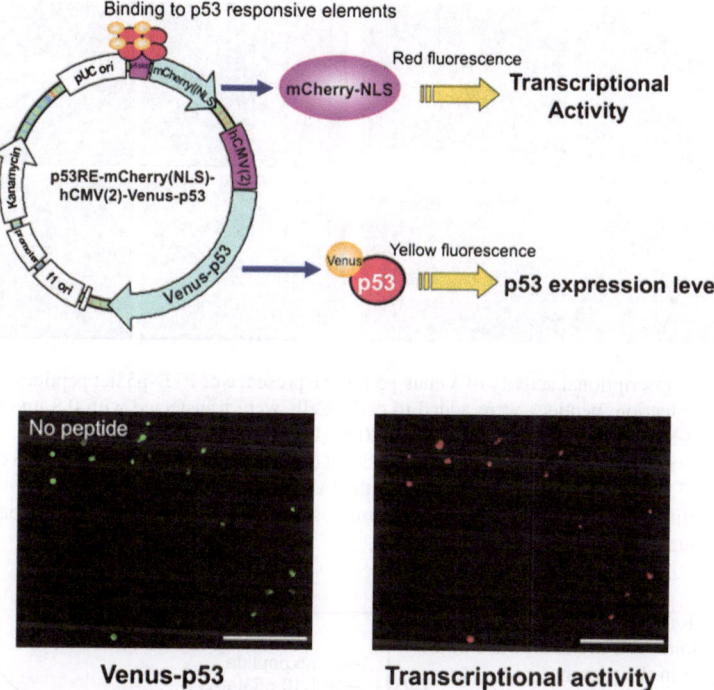

Fig. 4.5 Analysis of transcriptional activity of p53 protein. A p53 reporter system using yellow fluorescent protein for p53 levels and red fluorescent protein for transactivation by p53 was used. H1299 Cells were transfected with 0.8 μg of p53RE-mCherry-2-Venus-p53 in OPTI-MEM. After 1 h, RPMI 1640 medium with 10 % FBS was added and cells were cultured. After 11 h, the cells were fixed with 3.5 % formaldehyde and stained with DAPI. Venus (*green*) and mCherry (*red*) signals were quantified with a BZ-9000 florescence microscope (Keyence Corp., Osaka, Japan). *Scale bars* indicate 50 μm

Fig. 4.7, the p53 reporter assay showed that introduction of the K10- and R10-p53tet-WT peptides into cells significantly decreased the transcriptional activity of Venus-p53. In contrast, mutant K10- and R10-p53tet-L344P, which could not form tetramer and existed as unfolded monomer showed no effect on the transcriptional activity. These results suggested that the inhibitory activity of PTD-p53tet peptides was dependent on oligomerization. Hetero-oligomerization of Venus-p53 protein with PTD-p53tet peptide inhibited p53 transcriptional activity.

4.4 Discussion

p53 research has largely focused on the reactivation of mutant p53 tumor suppressor function. Inactivation of p53 leads to the development of cancer; therefore, the rescue of inactive mutant p53 function is important in cancer therapy.

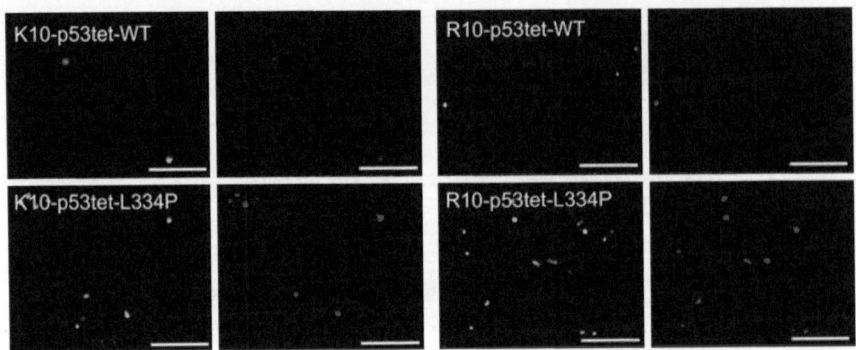

Fig. 4.6 Transcriptional activity of Venus-p53 in the presence of PTD-p53tet peptides. Six hours before transfection, peptides were added to cells. Cells were transfected with 0.8 μg of p53RE-mCherry-2-Venus-p53 using Lipofectamine 2000 (Invitrogen, USA) in OPTI-MEM. After 1 h, RPMI 1640 medium with 10 % FBS was added and cells were cultured. After 11 h, the cells were fixed with 3.5 % formaldehyde and stained with DAPI. Venus (*green*) and mCherry (*red*) signals were quantified with a BZ-9000 florescence microscope (Keyence Corp., Osaka, Japan). Scale bars indicate 50 μm

Fig. 4.7 Relative transcriptional activity of Venus-p53 in the presence of PTD-p53tet peptides (1 μM). The mCherry (*red*) signals in each cell, which expressed Venus-p53 at same level as endogenous p53 in A549 cells, were quantified. The averages of the mCherry signals in each EGFP signal are shown. *Solid circle solid line* no peptide; *open circle solid line* K10-p53tet-WT; *solid triangle solid line* K10-p53tet-L344P; *open triangle dotted line* R10-p53tet-WT; *solid inversed triangle dotted line* R10-p53tet-L344P

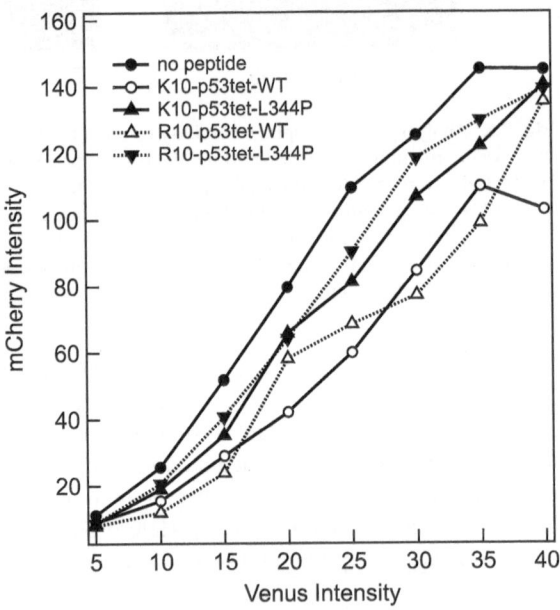

Recently, however, inactivation of p53 has received attention because heart failure is associated with up-regulation of p53 function. In addition, temporary suppression of p53 has been suggested as an approach to reduce the side effects of cancer treatment. Furthermore, it was reported that inhibition of p53 enhanced the efficiency of iPS cell generation.

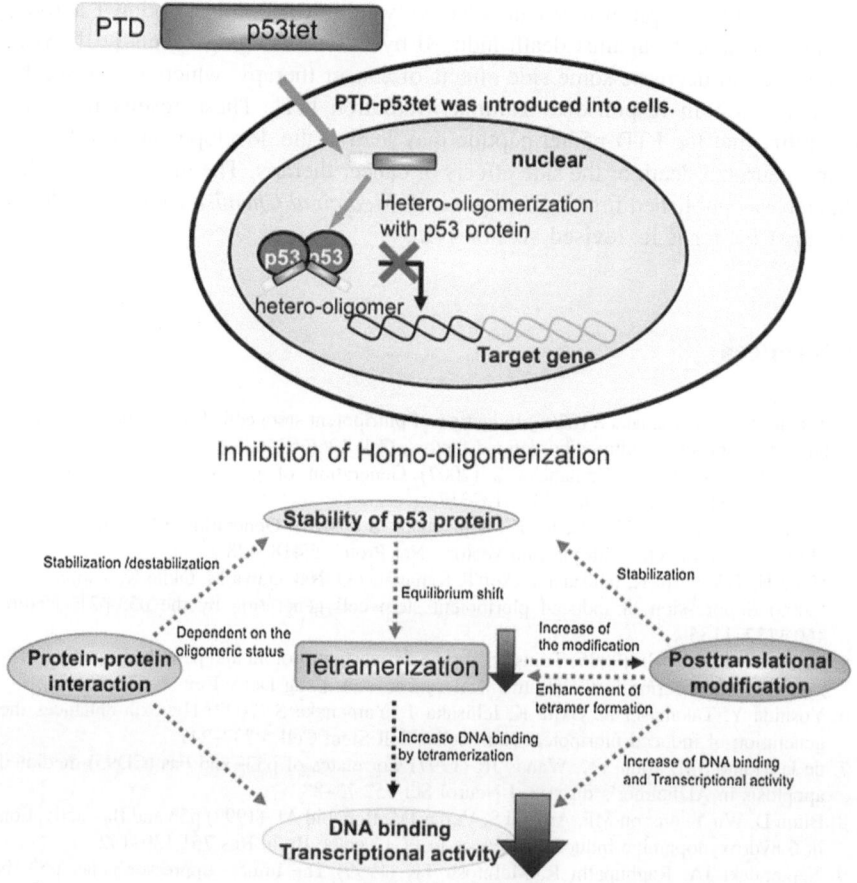

Fig. 4.8 p53 inhibition by PTD-p53tet peptide. Inhibition of homo-oligomerization of p53 leads to a decrease of p53 function

In this study, we demonstrated that introducing PTD-p53tet peptides into cells significantly decreased the transcriptional activity of p53 (Fig. 4.8). The inhibitory activity of the PTD-p53tet peptide was dependent on its ability to form a hetero-oligomer with the wild-type p53 tetramerization domain peptide, suggesting that the PTD-p53tet peptide could inhibit p53 transcriptional activity in cells through hetero-oligomer formation of the peptide with p53.

iPS cells exhibit functional and genetic properties similar to those of human ES cells. It has been reported that induced pluripotency can be achieved with as few as two pluripotency factors and can be enhanced by small molecules such as methylation inhibitors. Moreover, hypoxia enhanced the generation of iPS cells [6]. Inhibition of p53 function by the PTD-p53tet peptide might lead an enhancement in the efficiency of iPS cell generation. p53 has been implicated in the death of neurons in Alzheimer's disease, Parkinson's disease and traumatic brain injury [7–

9]. The compound pifithrin-α can selectively inhibit p53 transcriptional activity and protect neurons against death induced by DNA-damaging agents [10]. Also, pifithrin-α can decrease some side effects of cancer therapy, which are caused by p53 activation in response to gamma irradiation [11]. These results raise the possibility that the PTD-p53tet peptide may lead to the development of a drug to inhibit neuronal death or the side effects of cancer therapy. The main result in this Chapter was published first in *Bioorganic & Medicinal Chemistry Letters* in 2012, and this Chapter is its revised version [12].

References

1. Takahashi K, Yamanaka S (2006) Induction of pluripotent stem cells from mouse embryonic and adult fibroblast cultures by defined factors. Cell 126:663–676
2. Okita K, Ichisaka T, Yamanaka S (2007) Generation of germline-competent induced pluripotent stem cells. Nature 448:313–317
3. Okita K, Hong H, Takahashi K, Yamanaka S (2010) Generation of mouse-induced pluripotent stem cells with plasmid vectors. Nat Protoc 5:418–428
4. Hong H, Takahashi K, Ichisaka T, Aoi T, Kanagawa O, Nakagawa M, Okita K, Yamanaka S (2009) Suppression of induced pluripotent stem cell generation by the p53-p21. Nature 460:1132–1135
5. Wadia JS, Dowdy SF (2005) Transmembrane delivery of protein and peptide drugs by TAT-mediated transduction in the treatment of cancer. Adv Drug Deliv Rev 57:579–596
6. Yoshida Y, Takahashi K, Okita K, Ichisaka T, Yamanaka S (2009) Hypoxia enhances the generation of induced pluripotent stem cells. Cell Stem Cell 5:237–241
7. de la Monte SM, Sohn YK, Wands JR (1997) Correlates of p53- and Fas (CD95)-mediated apoptosis in Alzheimer's disease. J Neurol Sci 152:73–83
8. Blum D, Wu Y, Nissou MF, Arnaud S, Verna JM, Benabid AL (1997) p53 and Bax activation in 6-hydroxydopamine-induced apoptosis in PC12 cells. Brain Res 751:139–142
9. Napieralski JA, Raghupathi R, McIntosh TK (1999) The tumor-suppressor gene, p53, is induced in injured brain regions following experimental traumatic brain injury. Brain Res Mol Brain Res 71:78–86
10. Culmsee C, Zhu X, Yu QS, Chan SL, Camandola S, Guo Z, Greig NH, Mattson MP (2001) A synthetic inhibitor of p53 protects neurons against death induced by ischemic and excitotoxic insults, and amyloid beta-peptide. J Neurochem 77:220–228
11. Komarov PG, Komarova EA, Kondratov RV, Christov-Tselkov K, Coon JS, Chernov MV, Gudkov AV (1999) A chemical inhibitor of p53 that protects mice from the side effects of cancer therapy. Science 285:1733–1737
12. Wada J, Kamada R, Imagawa T, Chuman Y, Sakaguchi K (2012) Inhibition of tumor suppressor protein p53-dependent transcription by a tetramerization domain peptide via hetero-oligomerization. Bioorg Med Chem Lett 22:2780–2783

Chapter 5
Conclusion

Tetramer formation of p53 is essential for DNA binding, post-translational modification, and protein–protein interaction. To clarify the threshold for dysfunction of p53 in terms of the destabilization of p53's tetrameric structure, this study focused on the effects of tumor-associated mutations in the tetramerization domain on tetrameric structure and function. Furthermore, based on the structure–function analysis of mutant p53, I showed functional regulation of p53 via the control of tetramer formation.

In Chap. 2, by a comprehensive and quantitative analyses of 50 mutant tetramerization domain peptides, I showed that the effects of mutation on tetrameric structure were broad, and that the stability of the mutant peptides varied widely. The mutations in the hydrophobic core strongly destabilized the tetrameric structure, whereas the mutations at solvent-exposed residues had less of an effect on the tetrameric structure. At an endogenous p53 concentration in the nucleus, the fraction of tetramer for p53TD mutants was significantly decreased, even in the mutants that showed small destabilization effects in vitro. Furthermore, in addition to the direct effects of mutations on tetramer formation, several indirect effects may decrease p53 function via post-translational modifications and protein–protein interactions (Fig. 5.1). These observations suggest that the threshold for loss of tumor suppressor activity in terms of the disruption of p53's tetrameric structure could be extremely low.

In Chap. 3, I demonstrated that calixarene derivatives could increase the tetrameric stability of the mutant R337H, which is found in Li-Fraumeni syndrome, a hereditary disorder characterized by familial clusters of early-onset multiple tumors. Three calixarene derivatives, imidazole-calix[6]arene, pyrazole-calix[6]-arene, and guanidinium-calix[4]arene were designed and synthesized as stabilizers of mutant p53TD. The imidazole-calix[6]arene could stabilize the R337H mutant under physiological conditions and could enhance the transcriptional activity of p53-R337H in cells.

R. Kamada, *Tetramer Stability and Functional Regulation of Tumor Suppressor Protein p53*, Springer Theses, DOI: 10.1007/978-4-431-54135-6_5,
© Springer Japan 2012

Fig. 5.1 Correlation between tetramerization and function of p53. In addition to the direct effects of mutations on tetramer formation, several indirect effects may decrease p53 function via post-translational modifications, protein–protein interactions, and protein concentration

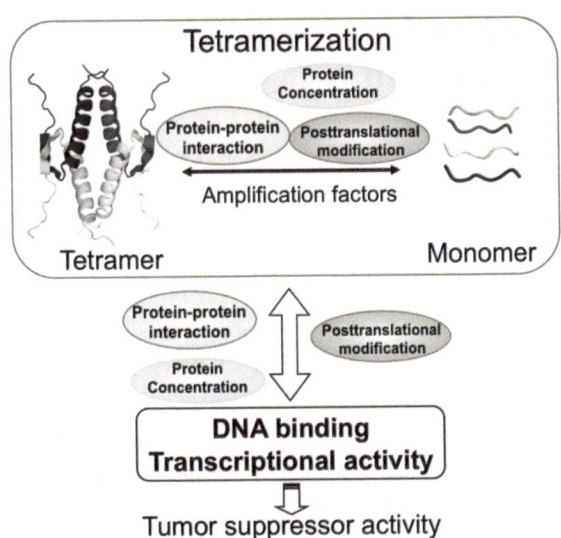

As shown in Chaps. 2 and 3, homo-tetramer formation if essential for p53 transcriptional activity. In Chap. 4, the inhibition of cellular p53 function via hetero-oligomerization with p53 was demonstrated. To introduce p53tet peptide into cells, a PTD was added to the N-terminus of p53tet. PTD-p53tet peptide could form hetero-oligomer with wild-type p53TD peptide in vitro. The p53 reporter assay revealed that the PTD-p53tet-wild-type peptide, but not the monomer mutant PTD-p53tet-L344P could significantly decrease the transcriptional activity of p53 in cells. These results suggest that the PTD-p53tet peptide inhibited p53 transcriptional activity in cells through hetero-oligomerization with p53.

I have demonstrated that tumor-associated mutations in the tetramerization domain destabilized the tetrameric structure and significantly decreased the tetramer fraction in the nucleus at endogenous p53 levels. Furthermore, the threshold for loss of tumor suppressor activity of p53 in terms of disruption of the tetrameric structure is suggested to be extremely low. In addition, a relatively small change of tetramer formation, by the stabilization or inhibition of hetero-tetramerization, could control p53 function. p53 suppresses tumor development through rapid activation in response to genotoxic stress. After inducing cell cycle arrest or apoptosis, p53 is quickly inactivated. This functional regulation by tetramer formation might be enabled in order to respond quickly to genotoxic stress (Fig. 5.1).